17th Pan-American Synchrotron Radiation Instrumentation Conference

(SRI 2013)

Journal of Physics: Conference Series Volume 493

Gaithersburg, Maryland, USA
19-21 June 2013

Editors:

Gwyn P. Williams

Peter Reversz
Uwe Arp

ISBN: 978-1-63266-718-2
ISSN: 1742-6588

17th Pan-American Synchrotron Radiation Instrumentation Conference

(SRI 2013)

Journal of Physics: Conference Series Volume 493

Gaithersburg, Maryland, USA
19-21 June 2013

ISBN: 978-1-63266-718-2
ISSN: 1742-6588

IOP Institute of Physics

Printed from e-media with permission by:

Curran Associates, Inc.
57 Morehouse Lane
Red Hook, NY 12571

Some format issues inherent in the e-media version may also appear in this print version.

Printed by Curran Associates, Inc. (2014)

For permission requests, please contact the Institute of Physics
at the address below.

Institute of Physics
Dirac House, Temple Back
Bristol BS1 6BE UK

Phone: 44 1 17 929 7481
Fax: 44 1 17 920 0979

techtracking@iop.org

Additional copies of this publication are available from:

Curran Associates, Inc.
57 Morehouse Lane
Red Hook, NY 12571 USA
Phone: 845-758-0400
Fax: 845-758-2634
Email: curran@proceedings.com
Web: www.proceedings.com

TABLE OF CONTENTS

ATOMIC AND MOLECULAR PHYSICS

BEAMLINE COMPONENTS

DETECTORS

FACILITY UPGRADES

FUTURE OPTICS

INFRARED AND TIMING

MICROSCOPY AND IMAGING

Journal of Physics: Conference Series **493** (2014) 011001 doi:10.1088/1742-6596/493/1/011001

Preface

17th Pan-American Synchrotron Radiation Instrumentation Conference SRI2013

These proceedings are a collection of the articles presented at the seventeenth Pan-American Synchrotron Radiation Instrumentation Conference SRI2013, held on the campus of the National Institute of Standards and Technology (NIST), located in Gaithersburg, Maryland, United States of America, 19–21 June, 2013.

SRI2013 was jointly hosted by the Cornell University Cornell High Energy Synchrotron Source (CHESS), the Thomas Jefferson National Accelerator Facility (Jefferson Lab), and the Synchrotron Ultraviolet Radiation Facility (SURF III) at NIST.

This meeting's focus was clearly on instrumentation, thus fulfilling the intent of this SRI meeting series, which was initiated at NIST, then the National Bureau of Standards (NBS), in 1979. SRI2013 hosted more than 150 delegates, despite the new US governmental travel restrictions. This proceedings series aims to be an essential reference work for practitioners in the field. It primarily documents the evolution and development of techniques, but also recent scientific advances, that were presented during the two and a half days of the conference. We are extremely thankful to all the authors who contributed to making these proceedings a volume of reference as well as to the reviewers for their careful reading and constructive recommendations for improving the articles.

Great thanks go to Robert Dragoset at NIST, for creating and maintaining the conference website and generating the conference logo. We are also thankful for the excellent support we received from the Conference Program at NIST, especially Kathy Kilmer and Angela Ellis. And we would like to dedicate these proceedings to the memory of Kathy Kilmer, who passed away on 15 October, 2013. NIST will not be the same without her.

The Co-Editors:

Uwe Arp (SURF/NIST)
Peter Reversz (CHESS)
Gwyn P Williams (Jefferson Lab)

Committees

Conference Chairs

Name	Organization
Gwyn Williams	Jefferson Lab
Peter Revesz	Cornell High Energy Synchrotron Source
Uwe Arp	Synchrotron Ultraviolet Radiation Facility

Programme Committee

Name	Organization
Alastair MacDowell	Advanced Light Source
Tom Toellner	Advanced Photon Source
Amitava D Roy	Center for Advanced Microstructures and Devices
Tom Ellis	Canadian Light Source
Roberta Santarosa	Laboratório Nacional de Luz Síncrotron
Jerry (Jerome) Hastings	Linac Coherent Light Source
Steven Hulbert	National Synchrotron Light Source
Thomas A Rabedeau	Stanford Synchrotron Radiation Lightsource
Mark Bissen	Synchrotron Radiation Center
Gwyn Williams	Jefferson Lab
Peter Revesz	Cornell High Energy Synchrotron Source
Uwe Arp	Synchrotron Ultraviolet Radiation Facility

Peer review statement

All papers published in this volume of *Journal of Physics: Conference Series* have been peer reviewed through processes administered by the proceedings Editors. Reviews were conducted by expert referees to the professional and scientific standards expected of a proceedings journal published by IOP Publishing.

40 years of Atomic Physics using Synchrotron Radiation

John West
STFC Daresbury Laboratory, Warrington WA4 4AD, UK

E-mail: john.west@stfc.ac.uk

Abstract Experiments in atomic physics using several synchrotron radiation sources worldwide over the last 40 years, and in which I have been personally involved, are described.

1 Introduction

It would be impossible for me to give a worldwide account of SR work in atomic physics, so in this paper I shall focus on some of the experiments in which I was involved and which marked my career in SR research. My story began in 1970, when I joined Geoff Marr's VUV Spectroscopy group at Reading University as a post-doctoral research assistant. I began to work closely with Keith Codling; at that time the main interest was in setting up a beamline on the Daresbury 5 GeV electron synchrotron to carry out photoionization measurements on gases. My task was to design and have built a grazing incidence monochromator with emphasis on order sorting ability, Throughout my career my primary interest and ability has been in the construction of optical instruments and experimental apparatus, with a view to putting them to use in making absolute measurements as well as measurements of electronic spectra.

2 Absolute photoionization cross section measurements of atoms: the heat pipe

The aim here was to measure the photoionization cross sections of metal atoms, and for this a vapour cell was required. The heat pipe was chosen, a device which contains the metal vapour by an inert buffer gas at each end, an internal mesh being used to circulate the condensed metal within the containment region. The technique avoids having to use vapour pressure curves, which are in many cases accurate to only a factor of two or worse, to determine the metal vapour density. By measuring the absorption difference between the cell with just buffer gas in it and the cell when hot, the absorption path length and the buffer gas pressure, the absolute cross section can be obtained. At this point it became clear that in the spectral region where we were working, 50 – 400Å, the helium buffer gas cross section was not known with sufficient accuracy. So we embarked on a series of measurements to determine this, expanding it to cover other rare gases.

The cell worked best at high temperatures, because that defined the length of the absorbing column better, and we noted that David Ederer and Caroline Mehlman were already using the same device to make measurements on lithium. Our work focused on sodium (1), which worked reasonably well, and cadmium for which it worked less well. Our measurements had an uncertainty of ± 25%, and for both these two atoms these remain, to my knowledge, the only absolute photoionization cross section measurements available in this spectral region.

3 Photoelectron spectroscopy of the rare gases

With the much greater photon intensity provided by synchrotron radiation compared to helium continuum or discharge sources, it became possible to make a continuous series of measurements of the partial cross sections and electron angular distributions on the inner shells of the rare gases. In this way electron-electron correlations could be highlighted, and the effect on the 5s and 5p electrons of xenon in the region of the 4d ionisation was seen, clearly reflected in the 5s and 5p cross sections and the electron angular distribution data (2), the latter being completely at variance with calculations at the time. Notably, a calculation using the RPAE method, by Miron Amusia and his student Vadim Ivanov (3) before they had seen the experimental data, was in excellent agreement with experiment.

4 Absolute cross section measurements on atomic ions

After the closure of the 5 GeV electron synchrotron in 1977, work on photoelectron spectroscopy (PES) of atoms and molecules ceased at Daresbury until the SRS came into operation during the early 1980s. During this period of shutdown I spent a year as SURF fellow at NBS, a year which proved to be highly enjoyable as well as very productive. Most of the research carried out during that period was on PES of molecules, focussing on shape resonant phenomena, so is not appropriate to this paper. However some atomic work was undertaken, on the β-values and branching ratios in the regions of the $3s3p^64p$ and $5s5p^66p$ $^1P_1$ resonances in argon and xenon (4). The sensitivity of the β-parameter to the resonances was evident; it was notable that the minima in β and the absorption cross section occur at different wavelengths

My interest in atomic work continued with measurements on the alkaline earths, in this case strontium, during time spent at the Photon Factory in Japan. Using a time-of-flight spectrometer, the single and double ionization spectra of the neutral atom were measured. But these were relative measurements; an entirely new kind of experiment in photoionization, brought to Daresbury by Ken Dolder and Barry Peart from the University of Newcastle in the UK, led the way to making absolute photoionization cross section measurements on atomic ions. They applied the technique of merged beam spectroscopy, which they had developed for beams of electrons and ions in their own laboratory, to make measurements on singly charged atomic ions. The absolute cross section can be readily calculated from this experiment; full details are given in reference 5. Initial results were not encouraging, since the photon flux from the 5-metre normal incidence MacPherson monochromator was much lower than expected. There was a hint of a high cross section around 580Å in the Ba^+ spectrum, but this result was already known from trial experiments using a helium lamp at the university in Newcastle.

It remained to find out the cause of the low photon flux from the monochromator, and this turned out to be due to some confusion over the direction of the blaze angle in the diffraction grating in the monochromator. Once this was corrected by rotating the grating in its mount, the flux increased 100-fold to 10^{12} photons/sec/Å. This, together with an improved method of aligning the photon beam to the ion beam, transformed the situation. The first results, on the Ba^+ ion, were staggering (5). We watched amazed as the signal at the main peak of the resonances in the 5p – 5d resonance region kept climbing, with a peak cross section of 3000Mb, ie 3×10^{-15} cm^2 and an oscillator strength of 2.8 for the four peaks seen. Further measurements on Ca^+ and Sr^+ followed in the region of their p-d resonances.

These data encouraged a great deal of theory activity, including a rare excursion on my part into theoretical work with Vadim Ivanov to try and identify the structure seen in the whole of the calcium spectrum in the photon energy region from 28 – 43eV. A later more complex calculation (6) confirmed and enlarged upon our earlier theoretical analysis.

At Daresbury there was no instrumentation available to continue such measurements at higher photon energies. A collaboration was begun with the University of Aarhus, using the grazing incidence monochromator originally used at Daresbury on the old NINA accelerator. An extensive programme of measurements on atomic ions was carried out on ASTRID using the same merged beam technique; details are contained in two reviews (7,8). I highlight here the interesting case of Cr^+, where it eventually turned out that theory was right and experiment wrong. The Cr^+ spectrum had been measured using the dual laser plasma technique in a collaboration between NIST and Dublin City University in both theory and experiment (9). It appeared that there was a remarkable difference between the Cr and Cr^+ spectra in the region of the 3p-3d and 3p-4d resonances, which theory could not reproduce. It was later realised that the experiment was probably affected by the presence of metastable states (10), a worrying problem with all ion spectroscopy measurements, and one which cannot always be solved. It was some years later before the experiment could be repeated using the merged beam technique at ASTRID (11). This allowed better control of metastable contamination in the case of Cr^+, due to the type of ion source used, and agreement with theory was much improved.

5 The "complete" experiment and double photoionization of metal atoms

In the last experiments to be presented in this paper, the aim initially was to measure all the photoionization parameters in the one experiment. This had already been done a while ago by Ulrich Heinzmann and his colleagues by measuring the spin polarisation of the ejected electron and its angular distribution; a different approach was adopted in this work.

The atom chosen was calcium, photoexcited in the 3p-3d resonance (12). One decay channel for this resonance leaves the Ca^+ ion in an excited 4p state after ejecting an electron. The ion then decays to the ground state emitting a photon, and by measuring the polarisation of the photon in coincidence with the photoejected electron, and in a separate measurement measuring the angular distribution of the electron, it is possible to determine the ratio of the dipole amplitudes of the two ejected photoelectron channels, and the phase difference between them. Absolute values of the dipole amplitudes could not be calculated since the total photoionization cross section of calcium was not known.

This experiment evolved into measurements of double photoionization of metal atoms, where the two ejected electrons were detected in coincidence and as a function of the angle between them, a triply differental cross section (TDCS) measurement. Calcium and strontium were chosen; by this time we had become experienced in producing vapour beams of these metals, and had developed an oven in which the light beam entered along the axis of the oven by means of a capillary light guide to maximise the interaction between the photons and the metal vapour. The two 127° analysers had to be protected from the vapour beam by trapping the vapour on a liquid nitrogen cooled shield. At the Daresbury SRS the photon flux was not sufficient to work off resonance, so we worked in the regions of the 3p-3d and 4p-4d resonances. Data were taken at 31.41eV, the peak of the 3p-3d resonance in Ca, for unequal energy sharing of the two photoejected electrons (13). In this case one of the analysers was fixed at 0°, ie along the main E-vector component of the incoming radiation. With the rotatable analyser at ± 90°, a value of 0.6 was measured for the photon polarisation, confirming that LS-coupling was valid at this photon energy. There were obvious differences between the two energy sharing cases. Secondary peaks were evident in one set of data, and also noticeable was the non-zero value of the TDCS when the two electrons were ejected in directly opposite directions, a breakdown of the symmetry rule which does not occur when the two electrons have equal energies.

Data were also taken for equal energy sharing of the two electrons. In this case the symmetry rule was observed, ie the TDCS was zero for the two electrons ejected at 180° to each other. The results were also different when the photon energy was changed to another position in the region of the 3p-3d resonance at a photon energy of 31.59eV where there is another peak in the photoionization spectrum. The peaks became broader although the symmetry rule was not broken. It is well known that in fact this resonance has many components, incompletely resolved in this experiment, and showed that the results obtained were definitely resonance dependent.

For this reason a further experiment was attempted at an off-resonance photon energy for calcium using the Trieste SR source ELETTRA, where an undulator beamline was available. Since the photoionization cross section was approximately 2000 times lower off-resonance, this was an extremely difficult experiment, even with the much improved intensity from the undulator beam-line. The fixed detector was placed as before aligned to the E-vector of the incoming beam, and the experiment was carried out for equal energy sharing of the two ejected electrons. The results (14) showed a marked difference from the corresponding case for helium (15), with the main peak occurring at a relative angle of 60° between the two analysers and being much narrower than the helium case

6 Acknowledgments

The above experiments have been chosen from the work in which I have been involved over 40 years, and perhaps represent milestones in my career over that period. What I have enjoyed most of all is the encouragement, help and enlightenment so willingly given by all those, worldwide, with whom I have worked. They are too many to mention individually here, so I present this paper as a tribute to them all.

7 References

1 Codling K, Hamley J R and West J B 1977 *J. Phys. B: At. Mol. Phys.* **10** 2797
2 Torop L, Morton J and West J B 1976 *J. Phys. B: At. Mol. Phys.* **9** 2035
3 Amusia M Ya and Ivanov V K 1976 *Physics Letters* A **59A** 194
4 Codling K et al 1980 *J. Phys. B: At. Mol. Phys.* **13** L693
5 Lyon I C et al 1986 *J. Phys. B: At. Mol. Phys.* **19** 4137
6 Hibbert A and Hansen J E 1999 *J. Phys. B. At. Mol. Opt. Phys.* **32** 4133
7 West J B 2001 *J. Phys. B: At. Mol. Opt. Phys.* **34** R45
8 Kjeldsen H 2006 *J. Phys. B: At. Mol. Opt. Phys.* **39** R325
9 Costello J T, Kennedy E T, Sonntag B F and Clark C W 1991 *Phys. Rev. A* **43** 1441
10 McGuinness C et al 1999 *J. Phys. B: At. Mol. Opt. Phys.* **32** L583
11 West J B 2003 et al *J. Phys. B: Atom. Molec. Phys.* **36** L327
12 Beyer H-J et al 1995 *J. Phys. B: Atom. Molec. Phys.* **28** L47
13 Ross K J et al 1999 *J. Phys. B: At. Mol. Opt. Phys.* **32** 2927
14 Beyer H-J et al 2000 *J. Phys. B: At. Mol. Opt. Phys.* **33** L767
15 Schwarzkopf O and Schmidt V 1995 *J. Phys. B: At. Mol. Opt. Phys.* **28** 2847

Footnote

This is a short report; an expanded version, complete with diagrams and a full set of references, is available by contacting the author using the e-mail address given above.

Journal of Physics: Conference Series **493** (2014) 012002 doi:10.1088/1742-6596/493/1/012002

Double photoionization of thiophene

R Wehlitz and T Hartman

Synchrotron Radiation Center, University of Wisconsin–Madison, Stoughton, Wisconsin 53589, USA

E-mail: rwehlitz@src.wisc.edu

Abstract. The emission of two electrons from the valence shell of an atom or molecule by a single photon is due to electron correlation. This so-called double photoionization process has been studied extensively for atoms and smaller molecules using different experimental techniques. Here, we discuss double ionization of aromatic hydrocarbons using monochromatized synchrotron radiation over a wide range of photon energies in combination with the ion time-of-flight technique. In particular we will discuss the case of thiophene (C_4H_4S), which is a molecule of pentagonal structure similar to furan and selenophene.

1. Introduction

Simultaneous excitation of two electrons in He, Ne and Ar has been investigated 50 years ago by Madden and Codling [1] in the first gas phase experiment using synchrotron radiation. Four years later, Carlson [2] reported the first observation of a "double electron ejection" from the outermost shell of an atom using an X-ray tube. These investigations initiated a host of experiments on double photoionization with synchrotron radiation, in which two electrons are excited into the continuum by a single photon. Naturally, He became the text book example for this process and many investigations using various techniques followed (see, e.g., [3]).

By the mid 1990s the ratio of doubly to singly charged He ions was settled and agreement between experiment and theory has been achieved. Investigations on Li, Be and Na demonstrated that the photon-energy dependence of that ratio is very similar to the one for He if the energy axis is in units of the energy difference between the corresponding double- and single-ionization thresholds [4]. Other elements, such as K and Mg, confirmed this observation, and even double K-shell photoionization exhibits the same energy dependence of the ratio as valence double photoionization. However, it was unclear whether this rule holds for molecules or not. An investigation on C_{60} demonstrated that the double-to-single photoionization ratio is affected by the structure of the molecule [5]. If the initially emitted photoelectron has a kinetic energy that corresponds to a de Broglie wavelength that matches a geometrical distance in the C_{60} molecule, so that a standing wave can be formed, then the electron stays longer in the molecule and increases thereby the probability for double ionization. Thus, the double-to-single photoionization ratio of C_{60} shows local maxima at certain energies that correspond to a de Broglie wavelength equal to the diameter of the molecule, or the diameter of a hexagon formed by the carbon atoms, or the distance between two adjacent carbon atoms.

The C_{60} experiment indicated that the structure of a molecule can affect the double-photoionization process. Therefore, we extended our research to aromatic hydrocarbons to see how the physical structure of a molecule affects the double-photoionization ratio. Experiments on acenes, chains of one to five benzene rings, demonstrated the existence of at least three different double-photoionization mechanisms [6]. One of them, already known from atoms, is the "knock-out" mechanism, in which the photoionized electron knocks out a second electron while leaving the molecule. The knock-out model has been introduced by Samson [7] based on the similarity of the excitation-energy dependence between the double-to-single photoionization ratio of an atom and the electron-impact cross section of the corresponding singly charged ion. In the case of the acenes, the ratio of doubly to singly charged parent ions rises linearly with the length of the molecule [8]. The second double-photoionization mechanism in acenes is the brief formation of a Cooper pair that breaks up into two individual electrons after photoemission [9]. This mechanism, however, does not exist for the aromatic molecules with a pentagonal structure such as pyrrole, furan and selenophene. The third mechanism observed for aromatic molecules leads to a linear increase of the ratio above a certain photon energy. At higher energies this ratio starts to increase even faster than linearly.

Here, we report of our results for thiophene (C_4H_4S), which is an aromatic molecule similar to furan and pyrrole. Therefore, we expected a similar photon-energy dependence of the ratio. However, thiophene exhibits a surprising and unexplained deviation from the typical photon-energy dependence.

2. Experiment

The experiment was performed on the 6-meter toroidal-grating monochromator (6m-TGM) beamline [10] at the Synchrotron Radiation Center (SRC) in Stoughton, WI (USA) using photon energies from 18 to 170 eV. Briefly, the monochromatic photon beam entered through a differential pumping stage into the interaction region inside the vacuum chamber, where the photon beam crossed an effusive beam of thiophene molecules. A pulsed electric field accelerated the photoions and photofragments towards a drift tube with a Z stack of microchannel plates at its end. More details regarding the experiment and data analysis can be found in [6]. We have detected doubly and singly charged molecular parent ions of thiophene (C_4H_4S) that are stable or long-lived metastable parent ions. While fragments are clearly visible in our ion spectra, they are not considered here.

3. Results and discussion

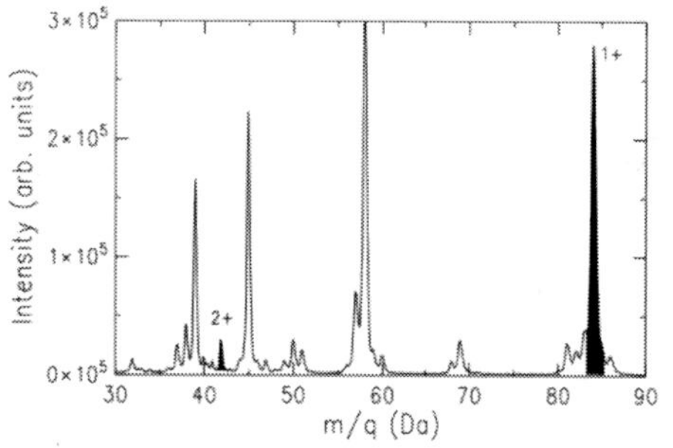

Figure 1. Ion time-of-flight spectrum of thiophene taken at 60 eV photon energy as a function of the mass-to-charge ratio. The doubly and singly charged parent ions are indicated.

Figure 1 shows an ion time-of-flight spectrum with the doubly and singly charged parent ions and fragments. Sulfur has four stable isotopes with the strongest one accounting for about 95%. In order to reduce errors we analyzed only the strongest peaks corresponding to the most abundent isotope. The peak of the doubly charged parent ions is well separated and its area was numerically integrated after background subtraction. The region near the singly charged parent ion (80-87 Da) exhibits the existence of other (much less abundent) isotopes as well as fragmentation (loss of a hydrogen atom). Therefore we determined the area of the singly charged parent ion containing the main sulfur isotope by a least-squares fit using a Gaussian profile.

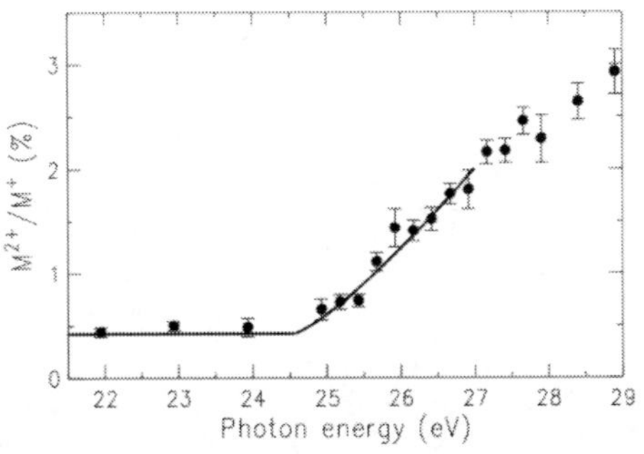

Figure 2. Ratio of doubly to singly charged parent ions of thiophene as a function of photon energy (full circles). The full line is a fit curve to the data (see text for details).

The ratio of the doubly to singly charged parent ions in the threshold region is shown in figure 2. We applied a least-squares fit to the near-threshold data using a power function:

$$R_s(\mathrm{h}\nu) = a(\mathrm{h}\nu - E^{2+})^n + b. \tag{1}$$

Here, $\mathrm{h}\nu$ is the photon energy, E^{2+} the threshold energy, a a fit parameter, n an exponent and b an additional offset. This empirical method to determine the double-ionization threshold has been used in the past and was referred to as the "nth power rule" [11] meaning an approximately parabolic fit for double ionization. A small offset b was included in the fit function to take care of any remaining doubly charged ion signal due to second-order light or scattered light. The threshold energy obtained for thiophene is (24.6 ± 0.5) eV, which is similar to the energies for pyrrole [24.20(6) eV], furan [25.18(14) eV] and selenophene [24.17(24) eV] [6].

In figure 3 we compare the ratios of doubly to singly charged parent ions of pyrrole [9], furan [9] and thiophene. The pyrrole data are divided by 1.73 and the furan data are multiplied by 1.5 for an easier comparison. At photon energies up to about 50 eV the ratios show a very similar photon-energy dependence that can be modelled by scaled He ratio data [12] shown as a gray broken line. This line represents the "knock-out" mechanism that contributes to the creation of doubly charged ions dominantly at lower energies but will continue to contribute also at higher energies [9].

At about 50 eV photon energy we see a linear increase of the ratio similar to the ratios for pyrrole and furan. The linear increase of the ratio extends over a rather large energy range of 35 eV for thiophene up to 100 eV for furan. For thiophene, however, the ratio decreases again above about 88 eV before the energy is high enough to photoionize a sulfur 2p electron near 165 eV when Auger processes start to contribute predominantly to the double-to-single photoionization ratio. The rather abrupt kink in the ratio of thiophene near 88 eV is surprising and cannot be explained right now. After subtracting the contribution to the ratio from the knock-out mechanism, we obtain the data represented by open circles in figure 3. These data indicate that the thiophene ratio is actually not decreasing but remains virtually constant up to about 150 eV where the S 2p threshold may start to affect the ratios.

4. Summary

Although the aromatic molecules pyrrole, furan and thiophene have a very similar (pentagonal) structure, the linear increase of thiophene's ratio does abruptly stop at 88 eV and remains essentially constant in contrast to the other molecules. This peculiar behavior of the thiophene ratio is not yet understood and will be explored in future experiments.

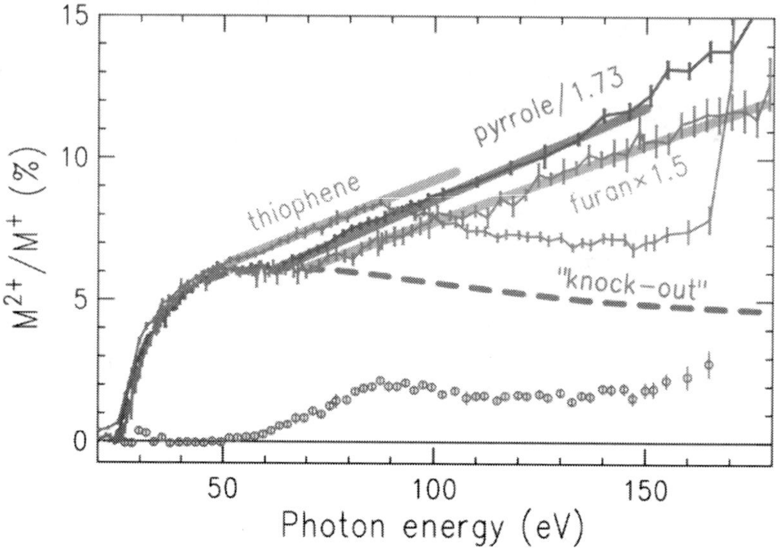

Figure 3. Ratios of doubly to singly charged parent ions of thiophene (red), pyrrole (purple)[9] and furan (green) [9] as a function of photon energy (data points with error bars connected by a full line). The ratios are scaled for easier comparison. The straight full lines emphasize the linear increase of the ratios for each molecule. The open circles are the thiophene data after subtracting the gray broken line.

Acknowledgments
The authors wish to thank the staff of the Synchrotron Radiation Center for their excellent technical support. We also thank Dr. Juranić for a critical reading of the manuscript. This work is based upon research conducted at the Synchrotron Radiation Center, University of Wisconsin–Madison, which was supported by the NSF under Grant No. DMR-0537588 and is currently supported mainly by the University of Wisconsin–Madison.

References

[1] Madden R P and Codling K 1963 *Phys. Rev. Lett.* **10** 516
[2] Carlson T A 1967 *Phys. Rev.* **156** 142
[3] Wehlitz R 2010 *Simultaneous Emission of Multiple Electrons from Atoms and Molecules using Synchrotron Radiation* (Advances in Atomic, Molecular, and Optical Physics vol 58) ed E Arimondo, P R Berman and C C Lin (New York: Academic Press) chapter 1 pp 1–76
[4] Bluett J B, Lukić D, Whitfield S B and Wehlitz R 2005 *Nucl. Instrum. Methods in Phys. Res.* B **241** 114
[5] Juranić P N, Lukić D, Barger K and Wehlitz R 2006 *Phys. Rev. Lett.* **96** 023001
[6] Hartman T, Juranić P N, Collins K, Reilly B, Makoutz E, Appathurai N and Wehlitz R 2013 *Phys. Rev.* A **87** 063403
[7] Samson J A R (1990) *Phys. Rev. Lett.* **65** 2861
[8] Hartman T, Juranić P N, Collins K, Reilly B, Appathurai N and Wehlitz R 2013 *Phys. Rev. Lett.* **108** 023001
[9] Wehlitz R, Juranić P N, Collins K, Reilly B, Makoutz E, Hartman T, Appathurai N and Whitfield S B (2012) *Phys. Rev. Lett.* **109** 193001
[10] Cole R K, Perkins F K, Brodsky E L, Filipponi A, Korpella E, Mancini D C, Pruett C H, Wallace D J, Welnak J T and Zanini F 1989 *Rev. Sci. Instrum.* **60** 2093
[11] Brehm B, Fröbe U and Neitzke H-P 1984 *Int. J. Mass Spectr. Ion Proc.* **57** 91
[12] Samson J A R, Stolte W C, He Z-X, Cutler J N, Lu Y and Bartlett R J (1998) *Phys. Rev.* A **57** 1906

Anisotropy and the optimal aspect ratio of a bent crystal

Mikhail Antimonov[1,2] **and Ali Khounsary**[2,1]

[1] University of Illinois at Chicago, Chicago, IL 60607, USA
[2] Argonne National Laboratory, Argonne, IL 60439, USA

E-mail: amk@aps.anl.gov

Abstract. A thin bent rectangular crystal plate can be used as the second crystal in a double-crystal monochromator to focus sagittally a wide synchrotron X-ray beam. When bent in the sagittal direction, the plate also bends in the tangential direction (so-called anticlastic bending), which can adversely affect the diffracted beam. It is known that the undesirable anticlastic deformation can be minimized if the plate aspect ratio (length to width) is optimized. The present study investigates the influence of crystal material anisotropy and shows that ignoring it in the design of a sagittally bent crystal can lead to a significant underestimation of anticlastic deformation. For example, analysis shows that the optimal aspect ratio for a simply supported bent Si(220) plate ranges from 2.13 to 2.82 depending on the in-plane orientation of the cut rectangular plate. Treating silicon as isotropic with material properties in the literature leads to a plate design with an optimal aspect ratio of 2.36 and a plate with significant anticlastic bending.

1. Introduction

Most synchrotron X-ray experiments can benefit from higher photon intensity that is often achieved by focusing a beam. Wide beams, such as radiation from bending magnet sources, are typically focused by sagittal focusing.

There are two common techniques for sagittal focusing a wide X-ray beam. One involves the use of cylindrical (or mildly conical) mirrors. Such mirrors provide achromatic focusing for relatively narrower x-ray beams with photon energy not much in excess of about 30 keV. At higher energies, the required incident angle becomes very small (< 3 mrad), leading to an elongated beam footprint and a rather long mirror [1] unless the mirror is multilayer coated. The rather small sagittal radii that are required, however, limit the optical aperture and the size of the beam the mirror can accommodate.

The second technique uses a sagittally bent second crystal in a double-crystal monochromator system, in either Bragg or Laue geometry [2-14]. Diffraction angles are much larger (even for higher-energy photons) compared with reflection angles off mirrors, allowing focusing of wider X-ray beams.

Sagittal bending, however, is accompanied with anticlastic (tangential) bending stemming from the non-zero Poisson's ratio of materials, as illustrated in Figure 1. While it can be exploited in the Laue geometry [2-4], anticlastic bending is generally undesirable in Bragg geometry, for it introduces a slight deviation to the incident angle ("slope error") resulting in an increase in divergence and a possible decrease in the diffracted beam intensity. Minimization of the anticlastic deformations within the beam footprint is often necessary (with the exception of situations where anticlastic bending is exploited for focusing [15]).

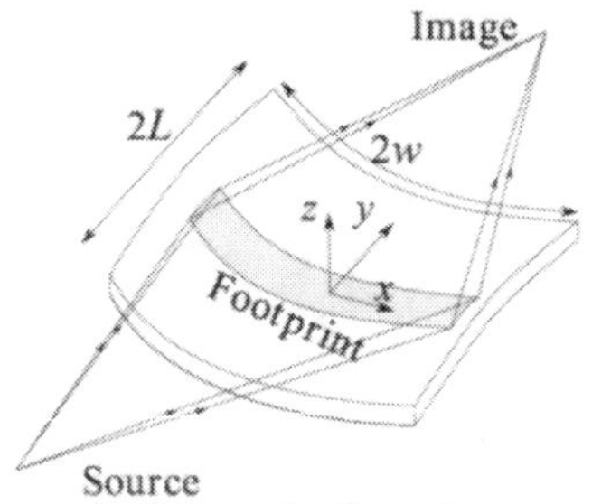

Figure 1. Schematic for a bent crystal plate. The filled area is the beam footprint. L and w are the half-length and half-width of the plate.

To reduce the anticlastic bending in diffracting crystals, several geometric remedies have been suggested, such as fabricating crystals with ribs [5,6] or with slots [7-10], or crystals in the form of thin, flat rectangular plates with an optimized length-to-width (aspect) ratio [11-14].

Crystals with ribs or slots can significantly reduce anticlastic curvature, rendering a large part of the crystal length usable. However, the crystals exhibit oscillations in the sagittal direction—due to the influence of ribs and slots—degrading focus quality [16]. These crystals are also costly to fabricate, are fragile, and cannot be generally bent to small sagittal radii without additional complexities [10].

Flat crystal plates, on the other hand, are not as fragile and are easy and inexpensive to fabricate. More importantly, their anticlastic bending can be minimized by an appropriate choice of crystal length-to-width ratio.

The dependence of anticlastic bending on the said aspect ratio was first realized by Kushnir et al. [11], who proposed a bent silicon crystal design based on the selection of an aspect ratio that results in a zero anticlastic curvature in the center of the bent plate. This optimal aspect ratio was shown to depend only on (1) the Poisson's ratio (silicon was assumed to be isotropic) and (2) how the plate was mounted. The optimal aspect ratio varies between 2 and 3.2 for simply supported isotropic plates and is less than 2 for clamped plates.

It was also shown [11] that there is an asymptotic solution: if the crystal is long (aspect ratio > 7), the anticlastic curvature could be minimized. Nisawa et al. [14] later showed that clamped plates with an aspect ratio greater than 3 would do likewise.

In these analyses, the plate material was considered to be isotropic, yet the crystal materials used for sagittal focusing are often anisotropic and can only be treated as isotropic in special cases. Neglecting anisotropy could result in a design with significant anticlastic deformations.

Krisch et al. [17] and Li [18] later showed that the anticlastic curvature in bent anisotropic crystals strongly depends on the crystallographic orientation. Their analysis was based on *beam* theory, which is valid only for narrow crystal slabs. The present paper concentrates on general *flat* plates. The dependence of the optimal aspect ratio on the crystallographic orientation of a crystal plate is shown. It is demonstrated that ignoring anisotropy in design can introduce significant tangential slope.

2. Optimal aspect ratio of a bent crystal plate

Let us consider a thin rectangular crystal plate (Figure 1) $2L \times 2w$ in size with the aspect ratio defined as $\gamma = L / w$. Plate edges at $x = \pm w$ are simply supported while the other two edges are free. A constant bending moment is applied to the supported edges to bend the crystal into a cylinder. It is assumed that the plate thickness is much smaller than other dimensions (for the thin plate theory to apply), and the maximum deflection is less than a quarter of the thickness (so the lateral forces can be ignored). With these assumptions, linear deflections depend on the applied moment. The ratio of anticlastic K_a to sagittal K_s curvatures everywhere in the bent plate can be shown [19] to be independent of the applied moment and can thus be used as the criterion for determining the optimal aspect ratio.

The analytical series solutions for bent plate profile were derived in [11,18] for isotropic materials and in [18] for anisotropic materials in specific orientations. The series converge very rapidly in the center of the plate where the anticlastic curvature must be zero at the optimal aspect ratio. Thus, the optimal aspect ratio can be derived from the first term in the series. For isotropic materials, the optimal aspect ratio $\gamma_{opt} = 2(3-\nu)/\pi\,(1-\nu)$, where ν is Poisson's ratio.

For anisotropic materials, the optimal aspect ratio also depends on the crystallographic orientation of the rectangular plate's diffracting surface as well as the orientation of its sides. A rectangular Si(220) plate, for example, can be cut from a (220) wafer in an infinite number of ways, so to specify the in-plane orientation of a rectangular plate, we introduce a 'cut angle' φ defined as the angle the long edge of the plate makes with a reference crystallographic orientation. Computation of the optimal aspect ratio is then carried out for various cut angles using the finite element (FE) method, because simple analytical solutions are only available for specific crystal lattice orientations [19].

The dependence of the optimal aspect ratio on a cut angle is illustrated through an example here. Consider a thin single-crystal Si(220) wafer from which a rectangular plate is cut such that its long side is parallel to the crystallographic direction ⟨001⟩. We denote this as having a cut angle $\varphi = 0$. The short side of the plate is thus parallel to the $\langle\bar{1}10\rangle$ direction.

For this primary in-plane orientation we use a dimensionless analytical solution [18] that provides the anticlastic-to-sagittal curvature ratio as a function of the aspect ratio for a rectangular plate bent into a radius $1/K_s$ for Si(220) with a cut angle of zero (Figure 2). Also shown is the solution for Si(111), which is in-plane, isotropic [11], and for which a cut angle need not be specified. The optimal aspect ratio is at $K_a/K_s = 0$. Similar curves for plates with any arbitrary cut angle can be constructed (using FE analysis) and their optimal aspect ratios determined. The result can then be presented in terms of the optimal aspect ratio as a function of the cut angle as presented in Figure 3 for Si (220). Also sketched in Figure 3 are three representative rectangular plates at different cut angles with different optimal aspect ratios. This gives the user some design options. For example, if the available (220) wafer has a small diameter, one would choose a 35° cut angle, which has the minimum aspect ratio requiring minimum plate length.

To demonstrate the impact of neglecting anisotropy, consider plate C in Figure 3. Let us assume that the crystal plate is 1 mm thick, 40 mm wide (thus, 112.8 mm long), and bent to a sagittal radius of 1 m. Figure 4 shows the computed slope of the anticlastic deformation in the central 20 mm of the plate (a reasonable value of a typical beam footprint) along the line that passes through the center (x=0 in Figure 1) of the plate. This slope is rather modest, as shown. However, if anisotropy is neglected and the plate is designed with an aspect ratio of 2.36 for isotropic silicon [11], the slope is increased more than 10 times.

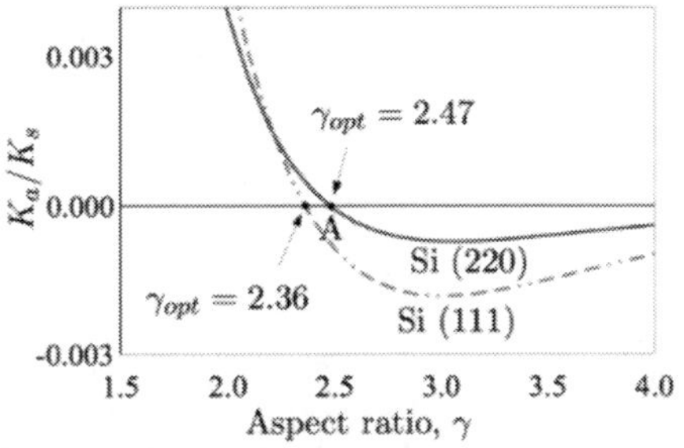

Figure 2. Ratio of anticlastic K_a to sagittal K_s curvatures versus aspect ratio for Si(111), which is in-plane isotropic and invariant in cut angle, and for Si(220) with a cut angle of zero (described in the text.)

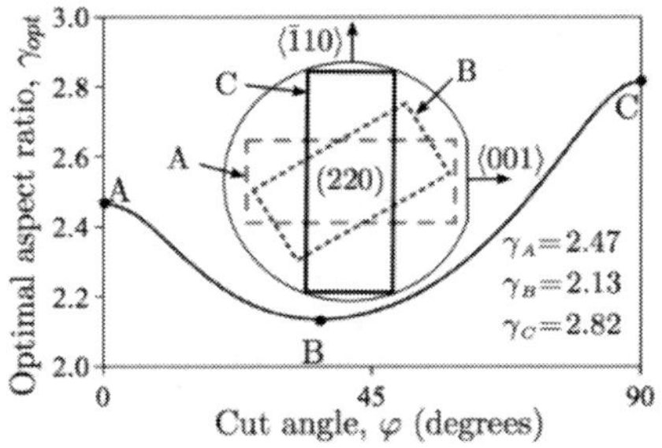

Figure 3. Dependence of optimal aspect ratio on the cut angle. A: initial in-plane orientation, $\varphi = 0°$; B: cut angle $\varphi = 35°$; C: cut angle $\varphi = 90°$.

3. Conclusions

Sagittally bent crystals with minimized anticlastic deformation provide an attractive option for x-ray focusing. The minimization is accomplished by incorporating stiffening structures in the crystal plate or by simply choosing rectangular plates with optimal aspect ratios that can be determined analytically or by finite element modeling. Since the crystals are anisotropic, in this paper the impact of material anisotropy on anticlastic bending and the determination of optimal aspect ratio were discussed, demonstrating the strong dependence of optimal aspect ratio (e.g., geometry of the crystal) on crystallographic orientation. The recipe for design consists of a few steps. First, based on the desired energy range and resolution, a crystal plane (and material) is selected. Then through finite element

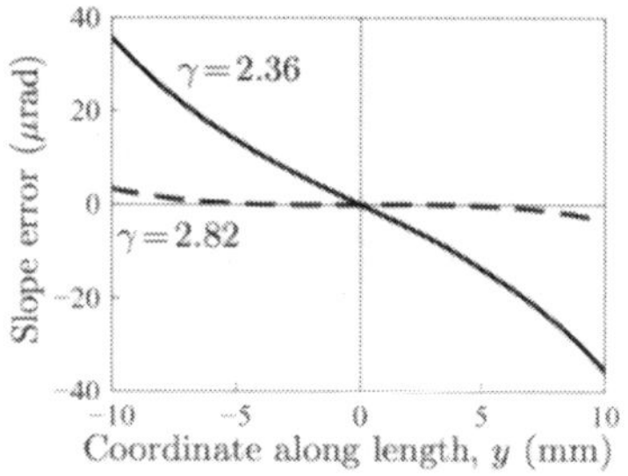

Figure 4. Tangential slope in the central part of a Si (220) plate (see text) with a cut angle of 90° (point C in Figure 3, γ_{opt} = 2.82). The slope for a non-optimal aspect ratio of 2.36 is also shown.

analysis (or analytical formulation, in certain cases), the variation of anticlastic-to-sagittal curvature ratio is computed and plotted against the plate aspect ratio. The aspect ratio at which the anticlastic curvature is zero is selected, and the process is repeated for a range of cut angles (as needed, based on the in-plane symmetry). A cut angle is then selected based on design considerations such as beam footprint size, available crystal size, etc., and the crystal is fabricated. The cut angle with the largest aspect ratio, for example, ensures both zero anticlastic curvature at the crystal center and the largest span of the crystal length with insignificant slope.

This analysis can be extended to asymmetric cut crystals where beam compression or expansion can be tailored to crystal tangential deformation.

Acknowledgment

This work is supported by the U.S. Department of Energy, Office of Science, under Contract No. DE-AC02-06CH11357. The encouragement and cooperation of Prof. F. Mashayek (UIC) and P. Strons (ANL) are graciously acknowledged.

References

[1] Susini J 1995 *Optical Engineering* **34**(2) 361
[2] Zhong Z, Kao C C, Siddons D P and Hastings J B 2001 *J. Appl. Cryst.* **34** 504
[3] Zhong Z, Kao C C, Siddons D P and Hastings J B 2001 *J. Appl. Cryst.* **34** 646
[4] Shi X, Ghose S, Zhong Z, Rehak M L, Kaznatcheev K, Takacs P Z and Dooryhee E 2011 *J. Appl. Cryst.* **44** 665
[5] Sparks C J, Ice G E, Wong J and Batterman B W 1982 *Nucl. Instrum. Methods* **194** 73
[6] Kheiker D M, Shishkov V A, Andriyanova M E and Smagin D V 2001 *Crystallography Reports* **46** 701
[7] Batterman B W and Berman L 1983 *Nucl. Instrum. Methods* **208** 327
[8] Yoneda Y, Matsumoto N, Furukawa Y and Ishikawa T 2001 *J. Synchrotron Rad.* **8** 18
[9] Lamble G M and Heald S M 1992 *Rev. Sci. Instrum.* **63** 880
[10] Kheiker D M, Shishkov V A, Shilin Y N, Rusakov A A, Dorovatovski P V and Zhavoronkov N V 2007 *Crystallography Reports* **52** 740
[11] Kushnir V I, Quintana J P and Georgopoulos P 1993 *Nucl. Instrum. Methods A* **328** 588
[12] Lang J C, Srajer G, Wang J and Lee P L 1999 *Rev. Sci. Instrum.* **70** 4457
[13] Feng L, Kang L, Li Z, Zhao F and Xu C 2008 *J. Synchrotron Rad.* **15** 140
[14] Nisawa A, Yoneda Y, Ueno G, Murakami H, Okajima Y, Yamamoto K, Senba Y, Uesugi K, Tanaka Y, Yamamoto M, Goto S and Ishikawa T 2013 *J Synchrotron Radiat.* **20** 219
[15] Shastri S (private communication 2013)
[16] Hazemann J L, Nayouf K and de Bergevin F 1995 *Nucl. Instrum. Methods B* **97** 547
[17] Krisch M, Freund A, Marot G and Zhang L 1991 *Nucl. Instrum. Methods A* **305** 208
[18] Li Y 2005 Ph.D. thesis Illinois Institute of Technology
[19] Antimonov M A and Khounsary A M 2014 Tech. rep. to be published as an APS Technical Bulletin (TB)

Upgrades to the XRD1 beamline optics and endstation at the LNLS

H Canova[1], A Fontoura[1], R T Neuenschwander[1], B Diaz[2] and C B Rodella[1]

[1]Brazilian Synchrotron Light Laboratory (LNLS) / Brazilian Center for research in Energy and Materials (CNPEM), C. P. 6192, 13083-970, Campinas, SP, Brazil
[2]Canadian Light Source Inc., 44 Innovation Boulevard, Saskatoon, SK S7N 2V3, Canada

E-mail: cristiane@lnls.br

Abstract. XRD1 was the first X-ray diffraction beamline to be built at the LNLS and after approximately 12 years of operation it was substantially updated to improve beam stability, increase the reliability of the monochromator movement as well as provide an experimental hutch that would meet the demands of users. The improvements included the construction of an independent concrete slab below the mirror and monochromator to minimize the vibrations originating from the floor. In addition, the installation of new monochromator mechanisms as well as the replacement of the two Si(111) crystals were performed in order to attain higher precision, stability and reproducibility during operation. Moreover, the diffractometer was replaced by a 3-circle heavy duty diffractometer from Newport to collect XRD patterns primarily in capillary geometry. A robotic arm was installed for fast and automated replacement of samples as well as to secure a cryojet or a hot air blower in front of the sample during measurements. In addition, a housing equipped with 24 Mythen detectors was installed at the beamline allowing for extremely fast data acquisition. Another upgrade was the integration of motors and control systems from PXI National Instruments and Galil controllers with Phytron. These systems are crucial for the next upgrade that is underway at the beamline: enabling remote access for users to collect their measurements without the need to travel to the LNLS.

1. Introduction

The Brazilian Synchrotron Light Laboratory (LNLS) is a 1.37 GeV machine and is the only synchrotron facility in Latin America [1]. It has been in operation since 1997 with the X-ray diffraction beamline (XRD1) being one of the first 7 beamlines initially built at the laboratory. The beamline is installed at the 1.7 T field B12 bending magnet and operates in the 5.5-14 KeV range [2]. Being the first diffraction beamline, the design of XRD1 was aimed at flexibility to accommodate a wide range of experiments including powder diffraction, reflectivity, monocrystal diffraction, multiple diffraction as well as inelastic scattering. In the following years two additional diffraction beamlines were built at the LNLS namely XRD2 (in 2003) and XPD (in 2005) with emphasis on monocrystal and powder diffraction respectively [3,4]. After approximately 12 years of operation the first X-ray diffraction beamline became uncompetitive compared to the more recent beamlines caused by lack of beam stability, low photon flux, precision and reproducibility of the monochromator motors as well as limitations to install furnaces and detectors in the diffractometer. Thus, during 2010 the XRD1 beamline was completely updated to improve beam stability, increase the reliability of the monochromator movement as well as provide an experimental hutch that would meet the requirements of users.

2. Optical enclosure upgrade

The optics of XRD1 beamline includes a vertically collimating/ cylindrically focusing bent Rh-coated mirror and a Si(111) double-crystal monochromator with developed at the LNLS. The second crystal of the monochromator is sagittally bent to focus the beam at the sample position producing a beam 3mm in the horizontal and 0.8mm in the vertical at 8 keV. Part of the beam vibrations originated due to the common floor slab supporting the optical components in the optical enclosure as well as the area outside the optical hutch. This problem was solved by building an independent concrete slab to support the mirror and monochromator. The concrete slab has a one inch gap to isolate the optical enclosure from the rest of the floor.

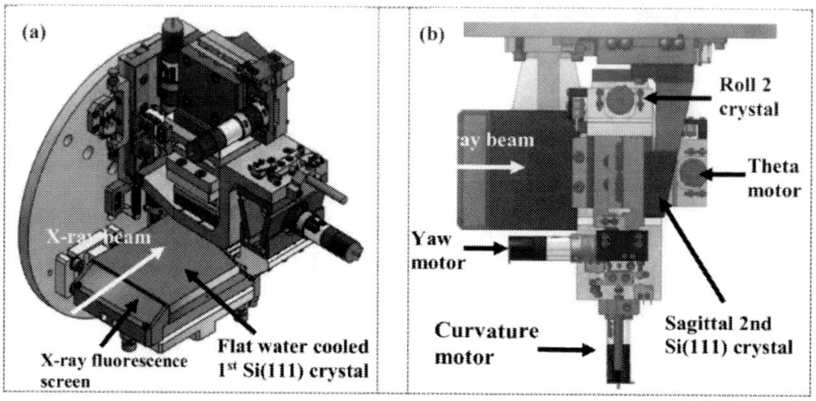

Figure 1. Monochromator mechanics design after the upgrade: (a) side view and (b) top view.

The second significant source of beam vibrations originated from the monochromator mechanics and the driver motors which wore out after many years of use. As a solution the monochromator mechanics and the support for the two crystals were completely redesigned. Figure 1 shows drawings of both the old and new design of the monochromator. The new mechanics are more robust than the old ones which now include titanium weak links, new motors and a new bender design. Further, a new Si(111) crystal from Crystal Scientific was installed to replace the first crystal. The thickness of this new crystal was increased by 5 mm to improve thermal stability. In addition, the thermal bath that is used to water cool the first crystal was also replaced by a LAUDA Proline RP845 to help avoid temperature fluctuations. Moreover, the control system of the monochromator was also renovated by the installation of a PXI National Instrument communicating with Galil controllers and Phytron ZMX+ electronics to power the stepper motor drives on an EPICS platform.

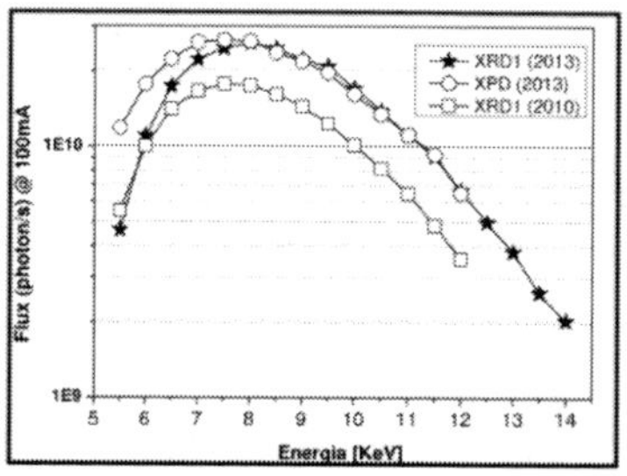

Figure 2. Photon flux as a function of energy: prior to upgrade (squares), after the upgrade (stars), and comparison with the powder diffraction beamline XPD at the LNLS (circles).

Following these upgrades in the first half of 2012 the commissioning tests confirmed significant improvements in the stability of the beam at the sample with vibrations below 100 μm after 10 h. Further, monochromator motor reproducibility and photon flux as a function of energy became comparable to the more modern powder beamline, XPD, at the LNLS (Fig. 2).

3. Experimental station upgrade

Nowadays the XRD1 beamline has two end stations, the first one was built in 1997 and included a 3 axis Huber diffractometer on a rotational stage to change the scattering plane from the vertical to the horizontal as well as an in-vacuum diffractometer. The second endstation was built in 2010 to house a thermo-mechanical simulation facility (Gleeble) [5]. The first endstation had several vacuum and mechanical problems that became more severe after several years of operation. The diffractometer also showed limitations regarding the load capacity for different sample environments and detectors. Based on the large number of scientific proposals submitted to characterize powder samples at the LNLS as well as the positive results regarding beam stability, precision and reproducibility of the motors during operation resulting from the optical enclosure upgrade, the decision was made to upgrade the first endstation of the XRD1 beamline to provide a high-throughput beamline for powder diffraction analysis. To accomplish this task the previous diffractometer was replaced by a 3-circle Heavy Duty diffractometer from Newport [6]. The diffractometer is suitable to collect XRD patterns in capillary geometry. A unique sample holder was developed in house to allow for flat-plate geometry. The Heavy Duty diffractometer has three coaxial high precision rotatory stages, one for the sample (Θ) and the other two for detectors (2Θ and δ) (Fig. 3). The considerable load capacity of the three circles (Θ = 35Kg; 2Θ = 100Kg and δ = 60kg) allows the installation of heavy sample holders or cell reactors and detector stages. Furthermore, to provide fast data collection, which is fundamental to rapidly characterize a large numbers of samples and to allow for time resolved experiments. This is achieved using an array of 24 Mythen detectors installed in the delta circle at a distance of 760mm from the sample and a Q resolution of the 0.02Å^{-1} .The Mythen detectors are placed inside a Huber built casing filled with helium (Fig. 3). In addition, a Yaskawa-Motoman robotic arm was installed for quick and automated replacement of samples as well as to hold a Cryojet 5 from Oxford Instruments allowing experiments from 100K to 500K as well as a soon to be acquired hot air blower from 298K -1273K.The next step of the XRD1 upgrade towards a high-throughput XRD beamline is the remote control of the endstation, which will allow access for users to collect their X-ray diffraction patterns without the need to travel to the LNLS. This is a work in progress and operation is foreseen to commence by the end of 2013.

Figure 3. XRD1 new end-station, with a Newport Heavy Duty diffractometer and a Yaskawa-Motoman robot for samples exchange and cryojet / hot air blower support.

4. Conclusions

The XRD1 beamline at the Brazilian synchrotron has undergone significant upgrades from the optics to the endstation. A noticeable improvement in beam stability and monochromator motor reproducibility was observed after the reforms in the optical enclosure. The experimental stations were replaced by a high-throughput modern powder diffraction station, reinforcing the industry focus of the beamline since the installation of the thermo-mechanical Gleeble instrument. The combination of a Heavy Duty Newport diffractometer with the 24K Mythen detector and robotic arm sample changer will allow for very fast measurements. These capabilities are desirable for *in situ* experiments as well as industrial applications. Future planned improvements for XRD1 include the implementation of remote access control of measurements for maximum process efficiency.

References

[1] Lira A C et al 1998 *Accel Conference Proceedings* 626-628
[2] Cusatis C, Kobayashi F M, Kakuno E, Giles C, Morelhao S, Mello V, Jr, and Mazzaro I 1998 *J. Synchrotron Rad.* **5** 491-493.
[3] Giles C, Yokaichiya F, Kycia S W, Sampaio L C, Ardiles-Saravia D C, Franco M K K, Neuenschwander R T 2003 *J. Synchrotron Rad.* **10** 430-434
[4] Ferreira F F, Granado E, Carvalho Jr W, Kycia S W, Bruno D, Droppa Jr R 2006 *J. Synchrotron Rad.* **13** 46-53
[5] J A. Ramirez, G. Faria, L. WU, C T. Alonso, A Isaac, J Piton, R.T. Neuenschwander 2012 *Proceedings VISUAL-JW 2012 Osaka Japan*
[6] Tang C C, Thompson S P, Hill T P, Wilkin G R, Wagner U W 2009 *Z. Kristalligr. Suppl.* **26** 153-158.

17th Pan-American Synchrotron Radiation Instrumentation Conference (SRI2013) IOP Publishing
Journal of Physics: Conference Series **493** (2014) 012005 doi:10.1088/1742-6596/493/1/012005

A proposal for a user-friendly modular framework for the simulation of x-ray experiments

G. J. Williams, A. Boehnlein, N. Graf, T. Koi, S. Xiao SLAC National Accelerator Laboratory, Menlo Park, CA 94025, USA

gjwillms@slac.stanford.edu

Abstract. As experimental concept and execution become increasingly complex across the spectrum of user-driven light sources, data analysis becomes more complicated and the urgency of completing an experiment on the first attempt grows. It has become apparent that simulation of the expected signal from such experiments can increase the likelihood of success of a particular experiment and the efficiency of a facility as a whole. Here, we propose a modular approach to the simulation problem, which will build on existing expertise and software. Additionally, we are working to support the simulation of partially coherent fields where the photon degeneracy parameter is much larger than unity.

1. Why invest in a new framework?

In a natural evolution of the x-ray user facilities and the communities they serve, state-of-the-art experiments increasingly rely on the detailed nature of the x-ray field interacting with the sample, whether this be related to the extremely large coherence volumes of current generation sources or the time structure of the probe in pump and probe experiments. Much of the optimization of these experiments is done "on the fly" during the allocated experiment time and is therefore not representative of the most efficient use of resources on the part of the facilities or their users. We believe that the excitement surrounding the rapid advances, especially in the spectral brightness and time structure of x-ray sources, and the increasing expertise of the user community can now be harnessed to provide a valuable toolset for simulating x-ray experiments and motivating even more exciting advances in sources (We note that projects similar to the one described here have been presented at previous international SRI meetings, c.f., the talk by Mancuso during Session P of the 2012 meeting.).

Many tools already exist for the simulation of the output of a particular beamline at a given source (SHADOW)[1] and Synchrotron Radiation Workshop (SRW)[2] are excellent examples of these tools.); however, in most cases these are readily usable only by experienced beamline personnel and, in our experience, the models are rarely up-to-date because of the rapid advances that typify the progress of current x-ray sources. In this paper, we propose to develop a framework that overlays these existing tools and provides an enhanced and flexible expert and end-user experience. As the project progresses, we anticipate the expansion of the framework to include the simulation of probe-sample interactions and mock data analysis, which will provide a "one-stop solution" for designing and optimizing x-ray experiments.

2. Details of the proposal

Arguably the most important part of the nascent framework is the basic structure of the user interface and the parameterization of the x-ray field. We believe that this process must involve a broad cross section of the community and continue to seek input from interested parties. We propose to discretize the simulation process by defining a series of discrete *events* that represent the propagation of a *data object* through vacuum or matter. For a schematic diagram, see figure 1.

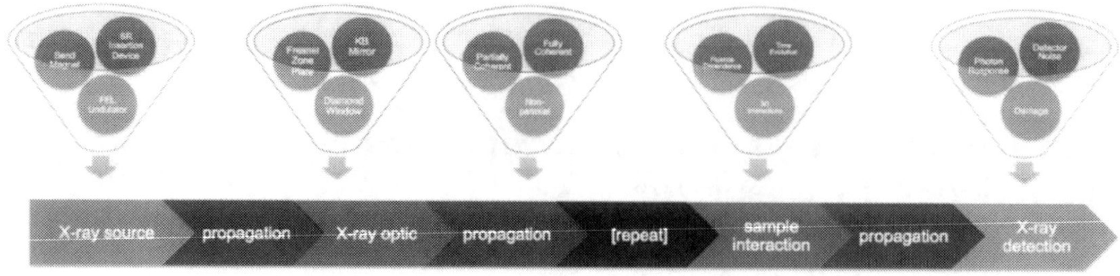

Figure 1. A modular, event-based approach to simulating an x-ray experiment from the source to the detector provides users the opportunity to use prebuilt models for each interaction or to provide their own, depending on the specific requirements of the experiment.

2.1. The data object

Our data object is the electric field. Provisions are made to store the following in a Cartesian space:

i. the complex field magnitude in the plane perpendicular to the propagation direction (We define this as the z-axis.), which is parallel to the Poynting vector at the source (We define this as the lab frame.) and subsequently determined by the field's interaction with matter;
ii. the components of the field's polarization;
iii. the energy spectrum of the field, defined around the mean frequency by a variable number of slices at user-defined resolution;
iv. the instantaneous time structure of the field, beginning at the "front" of the field at some time-zero and extending backward along the time axis;
v. inclusion of harmonic content, with the field's properties understood to scale as a multiple of the spectrum;
vi. the mean photon energy;
vii. the absolute energy that should be obtained by integrating the intensity over time, space, and frequency;
viii. the bounds on each of the field's variables(metadata);
ix. the increment of each of the field's variables(metadata);
x. the position in the 3D lab frame(metadata);
xi. the integrated optical path length(metadata) and total time of flight; and
xii. an Euler matrix describing the orientation of the current reference frame with respect to the lab frame(metadata).

2.2. The data structure

The data structure contains:

i. a multidimensional array with internal structure like so:
[(hor.,vert. polarization)][time][spectrum][(hor.,vert. position)][(real, imag. magnitude)];
ii. scalars representing the bounds and increments for the time, spectrum, and transverse position;
iii. scalars representing mean photon energy, absolute integrated energy, lab frame z-position, and optical path length;
iv. a vector indicating the pointing of the local frame's propagation direction relative to the lab frame;
v. strings containing unit information for the physical properties; and the identifier of the last module to handle the data object.

The object and its metadata are presented to a *module*—a method that manipulates the data object, e.g., propagation or a focusing mirror—and returns an updated data object.

2.3. Modules and events

Every *event* during the propagation of the simulated field is handled by a *module*. Every module accepts as input the data structure and metadata specific to that module, for example, a module that simulates the focused beam due to a pair of Kirkpatrick-Baez mirrors might take as metadata the x-ray incidence angle. Additionally, the module might include a detailed physical description of the field's interaction with the mirrors or a simple reflectivity curve.

Free space propagation is also treated in a module, where one can choose to propagate the field as a coherent object or take into account its detailed coherence properties. We are currently evaluating a coherent mode decomposition approach for the propagation of partially coherent fields[3]. For simulation of the field from the source to the sample location, we propose to remain in the paraxial approximation during free space propagation.

As the project matures, we anticipate the contribution of modules, especially those dealing with high field strengths or time dependent interactions, from the general user community.

2.4. The storage format

One HDF5[4] file is saved after the return of each module. The goal for the file format is two-fold: to record the information contained in the data structure and to record the "history" of the field. The file format has the following structure:

/
/event0
/<data object>
/<data object parameter 1>
.
.
.
/<data object parameter *N*=12>
/origination information (source identifier)
/author/time stamp
/event1
/data object
/...
/origination information (module identifier)
/author/time stamp
.
.
.

The /**event<*x*>** identifiers are either full data structures or symbolic links to groups within other HDF5 files.

2.5. Underlying tools

We believe that it is of vital importance to ensure that this framework is widely and freely available. To facilitate this goal and maintain a reasonable scope of work, we propose to embrace open source technology. In particular, we believe that the Python programming language[5] is an excellent foundation for the framework:

i. it is widely and freely available;

ii. modules written in nearly any programming language can be incorporated into the framework through the use of "wrappers;" and

iii. it natively supports "command line" input and is readily extended to include graphical user interfaces that are essentially platform independent.

3. Summary

We have briefly out-lined our current draft specification of a framework that we believe will enhance the user experience, planning activities, and efficiency of operation of x-ray facilities. We are eager to engage all invested parties in the development and realization of this project, while adhering to goals of open access and user contribution to the framework.

Appendix A. Acknowledgements

This work performed under DOE Contract DE-AC02-7600515. We thank the following individuals for encouragement and helpful comments concerning this project: J. Krzywinski, O. Tchoubar, Y. Chu, D. Shapiro, A. Mancuso, C. Yoon, L. Samoylova, J. Hastings, T. Rabedeau.

References

[1] Sanchez del Rio M, Canestrari N, Jiang F, and Cerrina F 2011 *J. Sync. Rad.* **18** 708.
The software website is: https://code.google.com/p/shadow3

[2] Chubar O, Fluerasu A, Berman L, Kaznatcheev K, and Weigart L 2011 *J. Phys.: Conf. Ser.* **425** 162001.
The software website is: https://github.com/ochubar/SRW

[3] Starikov A and Wolf E 1982 *JOSA* **72** 923.

[4] Folk M, Heber G, Kozoil Q, Pourmal E, and Robinson D 2011 *AD '11 Proceedings of the EDBT/ICDT 2011 Workshop on Array Databases* 36.

[5] See http://www.python.org for more information on the Python programming language.

Designing and commissioning of a prototype double Laue monochromator at CHESS

J Y Peter Ko[1,4], Benjamin B Oswald[1], James J Savino[1], Alan K Pauling[1], Aaron Lyndaker[1], Peter Revesz[1], Matthew P Miller[1,2] and Joel D Brock[1,3]

[1] Cornell High Energy Synchrotron Source, Cornell University, Ithaca, NY 14853, USA
[2] Sibley School of Mechanical and Aerospace Engineering, Cornell University, Ithaca, NY 14853, USA
[3] School of Applied and Engineering Physics, Cornell University, Ithaca, NY 14853, USA

peter.ko@cornell.edu

Abstract. High-energy X-rays are efficiently focused sagittally by a set of asymmetric Laue (transmission) crystals. We designed, built and commissioned a prototype double Laue monochromator ((111) reflection in Si(100)) optimized for high-energy X-rays (30-60 keV). Here, we report our design of novel prototype sagittal bender and highlight results from recent characterization experiments. The design of the bender combines the tuneable bending control afforded by previous leaf-spring designs with the stability and small size of a four-bar bender. The prototype monochromator focuses a 25 mm-wide white beam incident on the first monochromator crystal to a monochromatized 0.6 mm beam waist in the experimental station. Compared to the flux in the same focal spot with the Bragg crystal (without focusing), the prototype Laue monochromator delivered 85 times more at 30 keV.

1. Introduction

Sagittally-bent Laue monochromators have been in use to focus high-energy X-rays [1]. These monochromators make use of a set of asymmetric Laue crystals to focus the X-rays in the sagittal direction while taking advantage of the anticlastic bending in the meridional direction to minimize or eliminate the effect on the resulting energy resolution due to the vertical opening angle of the synchrotron beam [2].

We designed a prototype monochromator, which consists of sagittal bending devices (benders) for Laue crystals, slits, and translation and rotation stages. The prototype was commissioned on a bending magnet beamline at the Cornell High Energy Synchrotron Source (CHESS). Here, we show the design of the sagittal benders and highlight the results from the commissioning experiments.

[4] To whom any correspondence should be addressed

2. Prototype double Laue monochromator

2.1. *Prototype sagittal bender*

The primary goal of the design of the prototype sagittal bender was to be able to create a symmetric bend with reproducible sagittal bending radius (R_s) that will produce a focused beam in the experimental station, given the focal length of the beamline. R_s can be determined by:

$$R_s = \frac{4\sin\theta_B \sin\chi}{(1/F_1+1/F_2)}, \quad (1)$$

where θ_B is the Bragg angle and χ is the asymmetry angle, defined as the angle between the surface normal ad the diffraction plane [2], and F_1 and F_2 are source-to-monochromator and monochromator-to-experimental station distances, respectively. In addition to controlling the bending radius, twist adjustments in the crystals are necessary to avoid degradation of intensity and energy resolution of the resulting beam.

Two types of benders were considered: a leaf-spring, and a four-bar. A schematic representation of each type is shown in figure 1. In a leaf-spring bender (figure 1(a)), the crystal is clamped at its sides, and each side is mounted on a leaf spring. A sagittal bend is created by applying a force to one leaf spring while the other is fixed. Advantages of a leaf-spring bender are only a single motion is required to create a bend, and the design allows for a good thermal contact for heat exchange. On the other hand, because a force is applied to one side only, the bend is asymmetric – this becomes more problematic at higher energies, where the bending radius becomes smaller (see equation (1)). In a four-bar bender (figure 1(b)), a sagittal bend is created by displacing two inner bars, while the two outer bars remain fixed. With a four-bar bender, a symmetric bend can be created. Moreover, this design allows for a compact size of the bender. However, motions of the two inner bars need to be precisely synchronized, which adds complication to the design.

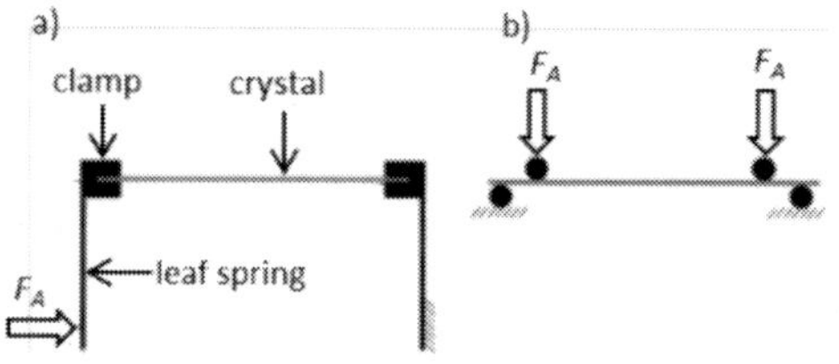

Figure 1. Schematics of a leaf-spring bender (a), and a four-bar bender (b)

Evaluating the two options, the four-bar bender was preferred, as compactness and being able to create a symmetric bend were key requirements. In our design, two sets of bending bars are used. Each set of bars is rotated opposite to one another to create a sagittal bend. Each bending bar has rotational freedom to allow the crystal to translate as it bends. A schematic representation of the bending mechanism is shown in figure 2(a). To create a bend, in principle, a moment can be applied to each side of the crystal with two synchronized motors; however, as previously suggested, this is one of the drawbacks of a four-bar bender. To overcome this issue, we implemented a differential screw design consisting of a dual-pitch rod. This rod contains two sections – one portion half the pitch of the second. A schematic representation of the design is shown in figure 2(b). One revolution of the motor shaft can move one point of application of force in the $+x$ direction, while the second point of application of force, mounted to the portion of the rod with twice the pitch moves in the $-2x$ direction, creating equal and opposite displacement.

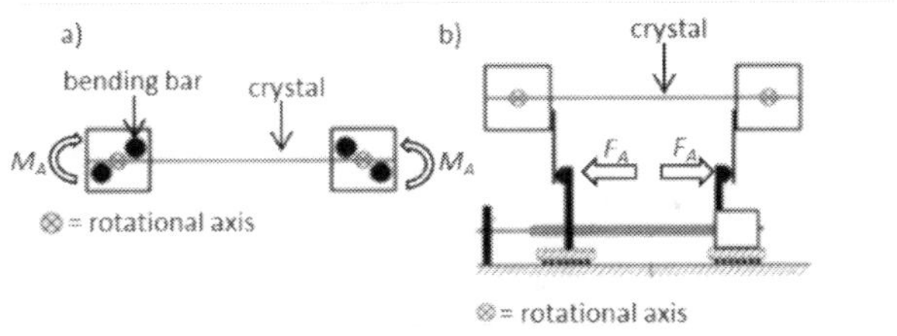

Figure 2. Schematics of the CHESS prototype four-bar bender showing the bending mechanism (a), and dual-pitch differential screw concept for a symmetric displacement with one motion (b).

Figure 3 shows the final design of the prototype bender. As described previously, a dual-pitch rod is used to actuate a bending motion. One revolution of the motor displaces the large-pitch threaded plate and the motor, each mounted to a translation stage containing a pushrod by the same magnitude in opposite directions. The motion of the pushrods is transferred to rotation in the sagittal cradles, each containing bender bars, through the sagittal extension plates. When the sagittal cradles rotate toward the crystal, the upper bars place compression on the top surface of the crystal, while the lower bars place tension on the bottom surface of the crystal. The bending radius of the crystal can be modified in approximately a 0.25 mm-increment. Rotational axis of each cradle is defined by a pair of sagittal bearings. One of the sagittal bearings is mounted on a twist cradle, which has its axis perpendicular to that of the sagittal cradle – this twist axis is defined by a single bearing.

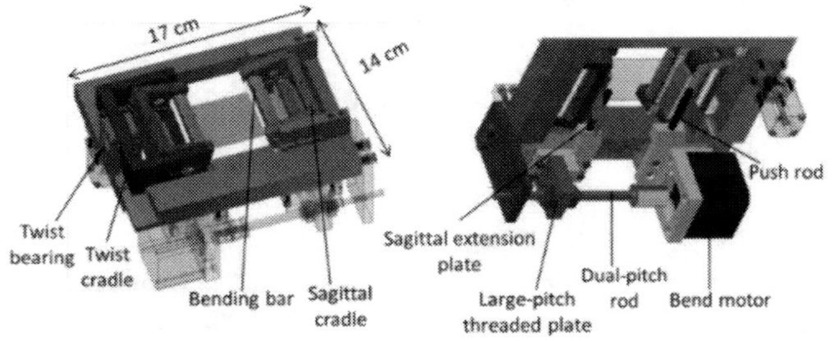

Figure 3. CAD models of the prototype bender viewed from top and bottom

2.2. Monochromator assembly

Figure 4 shows a model of the prototype Laue monochromator. The monochromator consists of a pair of sagittal benders (D). Each bender is mounted on a two-circle segment tilt stage (C), which in turn, is mounted on a rotation stage (B). The rotation stages provide the pitch motion, while the tilt stages provide the yaw and roll motions. Each rotation stage sits on a vertical (E) and a transverse translation (F) stages. In addition, the downstream assembly travels on a longitudinal translation stage (G), which provides the necessary separation between the upstream and downstream crystals necessary for energy tuning. Not shown in the model is a 5mm-thick aluminum filter, which was used to reduce heat load on the first monochromator crystal.

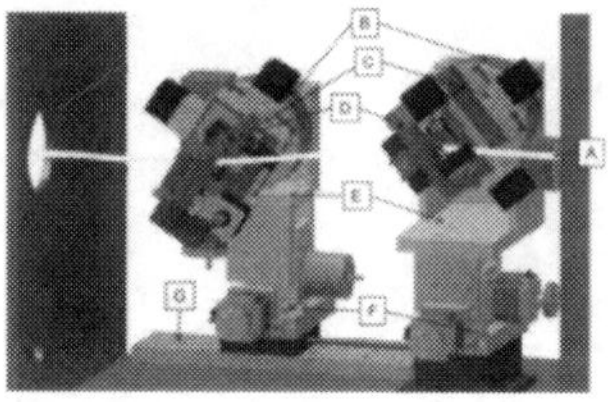

Figure 4. A CAD drawing of the prototype double Laue monochromator. A: X-ray beam (traveling from right to left), B: rotation stages, C: two-circle segment tilt stages, D: crystal benders, E: vertical translation stages, F: transverse translation stages, G: longitudinal translation stage.

3. Commissioning experiments

In the commissioning experiments, two 380 μm-thick 100 (h) × 40 (v) mm^2 Si(100) crystals were used. Each side was parallel to <110>, with the shorter side parallel to the bending axis. The experiments were performed at the B1/2 station at CHESS, which receives 5.9 GeV electron beam from a 0.55 T hard-bend magnet. The horizontal and vertical divergence was 2.5 mrad. The monochromator was located at 10 m from the source. Pitch and bending radius of the monochromator crystals were set to focus 30 keV X-rays (from the Si(111) plane) in the middle of the experimental station, 4 m from the monochromator. The bending radius was 435 mm. Flux of the incoming X-rays was monitored by a 6 cm-long ion N_2-filled chamber, and the beam on a phosphor screen was captured by a digital camera. Figure 5 shows an image of the beam at 4 m from the monochromator. Full-width at half maximum (FWHM) of the focused beam in the horizontal and vertical directions were 0.6 mm and 1.5 mm, respectively. FWHM of the beam on the first monochromator crystal in the horizontal direction was 25 mm. FWHM measurements were made with beam viewing software developed at CHESS. The measured flux in the focused beam was 3.9×10^{11} ph/s at 200 mA storage ring current, which corresponds to about 85× increases in flux with the Bragg crystals without focusing at the same energy (both measured and calculated); this increase is due to effects from both focusing and anticlastic bending. No degradation of beam quality was observed over several weeks.

Figure 5. An image of the focused beam in the experimental station, 4 m from the monochromator

4. Summary

We designed and commissioned a prototype double Laue monochromator at CHESS. The monochromator consists of several rotation and translation stage and benders. The design of the benders includes a use of a dual-pitch rod, which allows for synchronized movements of the bender bars with a single motor. The monochromator sagittally focuses a beam that was 25 mm wide at the monochromator, 10 m away from the source to 0.6 mm in the experimental station and 4 m from the monochromator.

5. Acknowledgements

CHESS is supported by the National Science Foundation (NSF) and the National Institute of Health/National Institute of General Medical Sciences under NSF award DMR-0936384. JYPK thanks financial support by the Energy Materials Center at Cornell, an Energy Frontier Research Center funded by the US Department of Energy, Office of Science, Office of Basic Energy Sciences under Award Number DE-SC0001086. We gratefully acknowledge Dr. Zhong Zhong of the Brookhaven National Laboratory for his guidance.

References

[1] Zhong Z, Kao C C, Siddons D P and Hastings J B 2001 *J. Appl. Cryst.* **34** 646
[2] Zhong Z, Kao C C, Siddons D P and Hastings J B 2001 *J. Appl. Cryst.* **34** 504

A fast white-beam shutter for hard x-ray topography at beamline 1-BM of the Advanced Photon Source

Naresh Kujala*, Mark Erdmann, Kurt Goetze, Joseph Sullivan, Xianrong Huang, and Albert Macrander

Advanced Photon source, Argonne National Laboratory, Lemont, IL 60439 USA

*kujala@aps.anl.gov

Abstract. Beamline 1-BM at the Advanced Photon Source (APS) delivers a white beam from a bending magnet with very intense x-ray photon flux. One important application of this beamline is white-beam x-ray topography imaging for crystal-based x-ray optics development and for industrial characterization of single crystals and epitaxial materials. Due to the intense photon flux from the third-generation synchrotron source of the APS, the exposure time of the imaging process should be accurately controlled down to the millisecond level. For this purpose we have designed and implemented a fast shutter that is vacuum compatible to 10^{-8} torr. The aperture is a copper block with a 70 mm horizontal and 5 mm vertical opening and is water cooled. The aperture is moved vertically up and down by means of a linear voice-coil actuator. The aperture's position is controlled using encoder feedback in a servo loop running on an industrial motion controller. A shutter opening response time of 32 milliseconds was measured. In this paper, we describe the shutter mechanics and its associated electronics installed at the 1-BM, and we report example white-beam topographs of diamond type IIa crystals.

1. Introduction

X-ray topography (XRT) is a very powerful tool for the evaluation of crystals for technological applications and for characterizing crystal and thin-film growth and processing. It is a technique capable of providing information on the nature and distribution of structural defects such as dislocations, inclusions/precipitates, stacking faults, growth sector boundaries, twins and low-angle grain boundaries in single crystal materials [1, 2]. In particular, synchrotron white beam x-ray topography is a fast and non-destructive characterization method for defects and of strain in bulk crystals, epitaxial layers, and electronic device structures [3, 4]. Synchrotron white beam x-ray topography is based on a Laue technique, that is, the diffraction Laue pattern consists of a number of large area diffraction spots, each one of which is a high resolution x-ray topograph. Typical exposure times for such topographs ranges from several tenths of a second to a few minutes. Owing to the intensity and brightness of the third-generation synchrotron sources such as the APS, the exposure time of the imaging process should be accurately controlled down to the millisecond level to avoid over exposure of the topographs. For this purpose we have designed and implemented a mechanical fast shutter that is vacuum compatible to ~10^{-8} torr. In this paper, we describe the fast shutter mechanics and its electronics and topographs collected using both fast shutter and personal safety shutter (PSS) of diamond sample are presented.

2. Shutter Design and Experimental setup

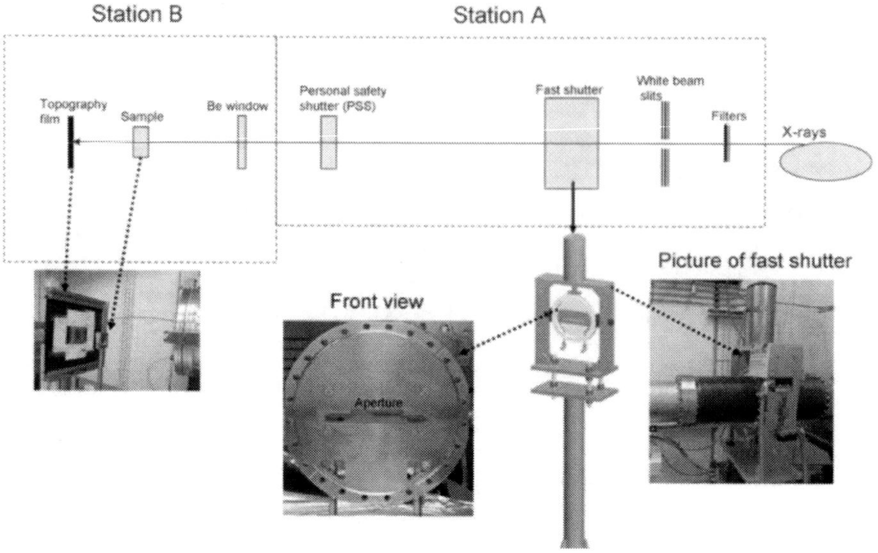

Figure 1: Beamline layout with inserted pictures of the fast shutter. The downstream film holder is also shown.

Beamline 1-BM at the Advanced Photon Source was recently reconfigured to an optics and detector testing beamline. The beamline shown in figure 1 consist of a first optics enclosure station A followed by two experimental hutches There is an upstream white beam B hutch and a downstream monochromatic beam C hutch. There are two modes of operations: white-beam and monochromatic-beam modes. A major program of the beamline is synchrotron white beam x-ray topography. X-ray optical components along the beam path that can be controlled by users and in the A hutch are first a white beam filters system, followed by white beam slit, then the fast shutter, and finally (not shown in the figure) a double crystal monochromator located in the A hutch. The fast shutter was installed in station A at 24 m from the source.

A copper block of fixed aperture was mounted on the 8 inch aluminum flange as shown in the figure 1. Both the upstream and downstream sides of the flange were connected to bellows which allowed actuation of the shutter, and the whole setup was mounted on a post-stand. The fast shutter was compatible with a high vacuum of $\sim 10^{-8}$ torr. The fast shutter has a fixed aperture with a horizontal opening of 70 mm and vertical opening of 5 mm. The aperture moves vertically up and down by using a linear voice-coil actuator. The linear voice coil actuator (BEI KIMCO MAGNETICS Company) is a single phase motor current rated for 20 amps (at 76 volts). It has a travel range of 15.8 mm with a peak force of 310 pounds. It has maximum allowable temperature of 155° C. The temperature rise in air per unit incident x-ray power is 0.85° C/Watt. We used incremental linear encoder (RENISHAW Company) with magnetic readout capable of 1 micron positional accuracy. The position of the aperture was controlled using encoder feedback in a servo loop running on an industrial motion controller. A Gemini GV6K-U12E digital servo driver (Parker Company) was used to control the position, speed, and acceleration of the fast shutter motion. This digital servo driver has 12 Amps peak current, 3.5 kW of peak power and is configurable via Ethernet or RS232 ports. A thermocouple was mounted on the voice coil actuator to monitor the temperature (DP-41 Omega). The position and speed of the fast shutter motion are controlled using EPICS (Experimental Physics and Industrial Control System) software.

3. Results

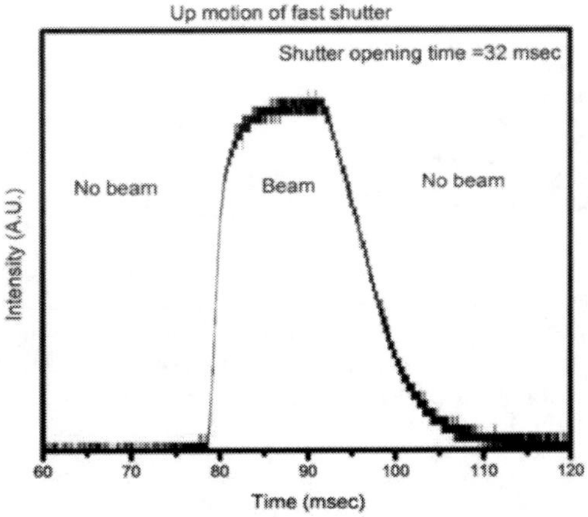

Figure 2: Measured white beam performance of the shutter. The data was collected by placing a scatterer in the beam path. The data in the figure is from a pin diode detector when the shutter was moved vertically in upward directions.

Timing tests of the fast shutter were performed with a scatterer and a pin diode detector placed in station B. The size of the beam was controlled by white beam slits and for the timing measurements the slits openings were 5 mm (H) and 3 mm (V). We used an aluminum block as scatterer and placed it in the beam path at 45° to the x-ray beam. The scattered beam was collected by a pin diode (with a 28mm x 28mm opening window) which was placed 15 cm from the aluminum block. A digital oscilloscope (4-channels 1 GHz – 10 GSample/s, RTO1014 Rohde and Schwarz) was used to measure the signal from the detector in self triggering mode. The opening and closing of shutter was controlled by the EPICS software from the computer outside the hutch. Figure 2 shows the measured time response motion of the fast shutter. The data were collected by moving the shutter up and down motion. A shutter opening response time of 32 milliseconds was measured.

In order to demonstrate its intended performance, white beam x-topographs of a diamond crystal type IIa sample were obtained with the newly installed fast shutter as well as with the much slower personal safety system (PSS) shutter which could be controlled to only ~1 sec opening time. The experimental setup is shown in the figure 1. Ozone levels were monitored (model: IN2000-L2-LC, from IN USA INCORPORATED) regularly during the experiment. The x-ray film was 8 inch x 11 inch in size and was mounted on a film holder as shown in figure 1.The distance between the sample and x-ray film was 12 cm. Transmission Laue geometry was used to take the white-beam topographs using both the regular PSS and the white-beam fast shutter. The topographs are shown in figure 3 and were indexed with the LauePt software described in Ref.5. As shown in figure 4a, the Laue patterns recorded using the PSS with an exposure time of ~1 second, the fastest time available with the PSS shutter, revealed that some of the Laue spots were over-exposed. The same Laue spot recorded using the fast shutter with an exposure time of 32 milliseconds is shown in figure 4b. Much reduced background darkening was revealed, which should be more amenable to further analyses, e.g. for studies of weak fringe systems as often occur in topography of stacking faults [6].The Laue spots shown in figure 3 were magnified using Nikon microscope (Nikon eclipse LV100POL) to produce the images shown in figure 4.

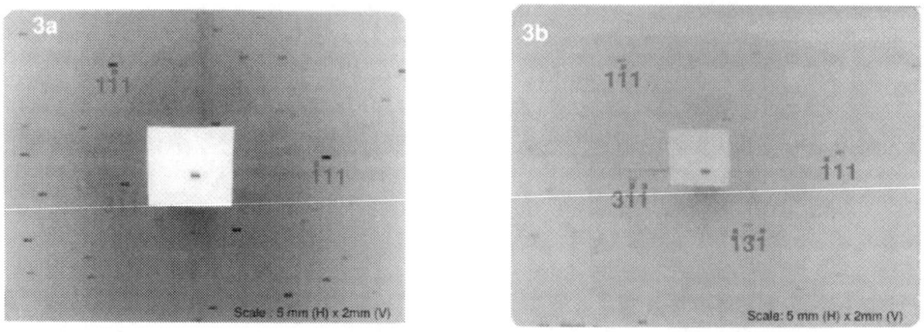

Figure 3: White-beam transmission Laue patterns of a (111) type IIa diamond crystal recorded on 8 inch x 11 inch X-ray film; each diffraction spot is a topograph of the crystal (see Fig. 4). 3a: Laue pattern taken with the personal safety shutter (PSS) with opening time of ~1 seconds, where many diffraction spots are too dark due to over exposure. 3b: Laue pattern taken with the white-beam fast shutter with opening time of 32 milli-seconds (accurately controllable from 30 milli-seconds to a few seconds).

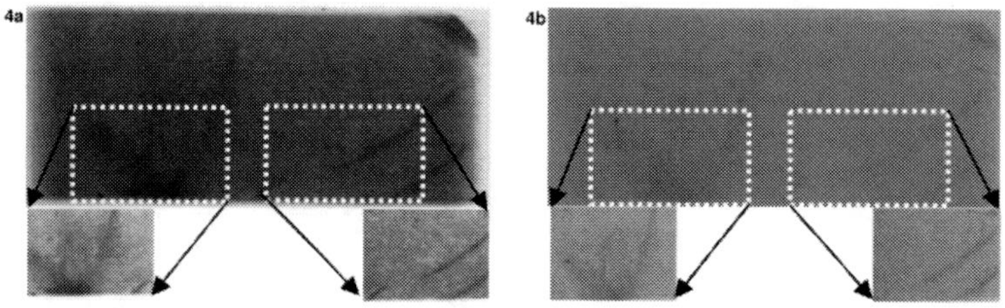

Figure 4: Topographs of the type IIa diamond crystal magnified from Fig. 3; 4a: $\bar{1}11$ reflection topograph taken with the PSS with opening time of ~1 second. 4b: $\bar{1}11$ reflection topograph with the fast shutter with opening time 32 milli-seconds, showing finer and weaker contrast features.

4. Conclusion

We have installed and tested a hard x-ray white beam fast shutter at the optics and detector 1-BM beamline of the Advanced Photon Source. The fast shutter can be controllably opened down to a 32 msec opening time. Topographs for a diamond crystal were compared to those obtained with a personal safety shutter (PSS) that could operate no faster than ~ 1 sec and the fast shutter resulted in much improved background exposure.

Acknowledgements: We would like to thank Linda Young, Mark Beno, Jonathan Lang, Chris Jacobsen and Lahsen Assoufid, and Advanced Photon Source Engineering Support Division (AES) staff for their support. This work was supported by the U.S. Department of Energy, Office of Science, under contract No. DE-AC-02-06CH11357.

References

[1] M. Dudley, and X. R. Huang. *MRS Proceedings,* **307**, (1993).
[2] B. Raghothamachar, G. Dhanaraj, J. Bai, and M. Dudley. *Microsc. Res. Tech.,***69,** 343–358, (2006).
[3] A. N. Danilewsky, J. Wittge, T. Weitkamp, R. Simon, H. Riesemeier, and T. Baumbach. *Nucl. Instrum. Methods Phys. Res. B,* **266**, 2035-2040 (2008).
[4] M. Moore, *Crystallograpy Reviews* **18(3)**, 207-235. (2012).
[5] X. R. Huang, *J. Appl. Cryst.* **43**, 926-928 (2010).
[6] A. Authier, "Dynamical Theory of X-ray Diffraction", *Oxford Univ. Press*, 2001.

Journal of Physics: Conference Series **493** (2014) 012008 doi:10.1088/1742-6596/493/1/012008

50 MHz Voltage-to-Frequency Converter

T Madden, J Baldwin

Advanced Photon Source, 9700 S. Cass Ave.
Argonne National Laboratory, Argonne, IL, 60439.

tmadden@aps.anl.gov

Abstract. The Voltage-to-Frequency Converter (V2F) is an instrument commonly used in synchrotron beam line experiments for converting a slowly varying voltage signal representing x-ray flux to a high frequency pulse train, with each pulse representing a packet of x-ray energy. The pulses are commonly integrated with a digital counter to calculate total x-ray energy incident on a sample during an exposure. As x-ray experiment data is taken faster and faster and exposure times shrink, V2F frequencies must necessarily increase to preserve precision in measuring total x-ray energy. The APS Detector Group have designed and built two custom 50MHz V2F instruments for APS beam line experiments. The instruments are designed according to the Nuclear Instrument Module (NIM) standard and respectively reside at APS Sector 12-BM, and in the APS Detector Pool [1].

1. Introduction

Voltage-to-Frequency Converters (V2F) are ubiquitous instruments at synchrotron experiments. The instruments, as their name implies, convert a varying voltage into a series of pulses at a frequency commensurate with the voltage. A common use of the V2F in a synchrotron experiment is to convert a voltage output from an x-ray detector, such as an ion chamber, into a pulse train to trigger a scalar counter. Many detectors, such as ion chambers, are sensitive to x-ray flux and do not respond to individual photons. The flux measured from an ion chamber is proportional to x-ray power delivered to the sample, that is, energy per unit time. It is desirable to integrate that power to determine total deposited x-ray energy on the sample. One may ask the question on why to use a V2F when all is needed are a modern analog-to-digital converter (ADC) and digital integrator. As will be seen, the V2F is exactly this circuit: an ADC and digital integrator. The reason for using a separate V2F unit and scalar is that a single data acquisition system can be used for counting detectors, such as photomultiplier tubes, or flux sensitive detectors such as ion chambers. The V2F makes a flux sensitive detector produce an output similar to a counting detector.

Many beam lines, such as Sector 12-BM at the Advanced Photon Source, possess a commercial V2F from vendors such as Nova R&D [2]. These commercial units have a frequency range from 0Hz to 2MHz, with a voltage input from 0V to +/- 10V. For most measurements, these commercial instruments suffice for practicing good science. However, as detectors get faster and faster, and exposure times get smaller and smaller, faster V2F converters are necessary to preserve instrument resolution. To illustrate, for a 10ms exposure and 2MHz frequency, the V2F will deliver 20000 counts. Taking log2, this comes to almost 15bit resolution, which is fine in most cases. If the exposure time is

1ms, the 2MHz V2F delivers 2000 counts, and hence has 11 bits of resolution. One can see that for shorter and shorter exposure times, the resolution of the V2F gets worse. If the V2F delivers a 50MHz frequency, we have almost 16bits resolution for a 1ms exposure. This allows precise measurement of energy deposited on a sample for short exposure times, speeding up the process of data gathering, and reducing sample damage.

2. Algorithm for V2F

The APS detector group designed a custom V2F circuit using a commercial high-quality analog to digital converter (ADC) and field programmable gate array (FPGA). The voltage fed into the APS V2F instrument is first processed with an analog amplifier, and then digitized with the ADC. The digital signal is sent to the FPGA which generates a pulse train based on the digitized voltage. A diagram of the V2F algorithm is shown in figure 1.

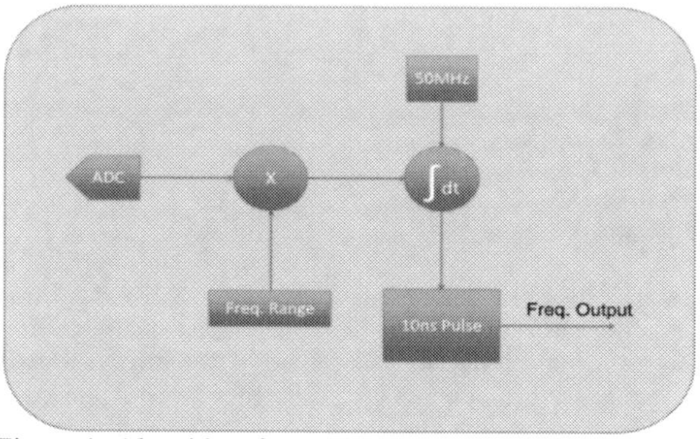

Figure 1. Algorithm for APS V2F convertor. Digitized input voltage is converted to pulse train in FPGA.

The digitized voltage is multiplied by a constant determining the output frequency range. This product is considered to be a fraction, that is, a number between 0.0 and 1.0, represented by a 32 bit binary number. The product is then integrated in a digital integrator clocked by a 50MHz system clock. When the running sum in the digital integrator reaches 1.0, the integrator overflows. On overflow, the integrator drops the integer part of its value. For example, if the sum of the integrator reaches 1.1, then the 1 is subtracted and the result stored in the integrator is 0.1. Each overflow causes a 10ns pulse to be generated. The 10ns pulse is timed by an internally generated 100MHz clock in the FPGA. The series of these pulses comprises the frequency output of the instrument.

The beauty of this algorithm is its simplicity and its ability to generate frequencies only limited by the system clock. For example, using a 100MHz system clock would allow the V2F to generate 100MHz frequencies with the same performance in terms of linearity and accuracy, as these parameters are limited by the quality of the input amplifier and ADC. This algorithm was based on an invention described in [3,4].

3. The APS Instrument

3.1. High Level Description of Instrument

The APS V2F converter is built as a NIM instrument because virtually every APS beam line has NIM crates. The NIM power supplies are low noise, and lend themselves to powering analog instrumentation. The APS instrument features four V2F channels, with four voltage inputs and four frequency outputs, all independently configurable. The inputs and outputs on the instrument appear as BNC connectors on the front panel. The instrument is configured with a knob and pushbutton switch that control a series of menus displayed on a liquid crystal display (LCD). Each channel can be configured to five frequency ranges: 0MHz (disabled), 2MHz, 10MHz, 25MHz, and 50MHz. The voltage input ranges from 0V to +10V, corresponding to 0MHz to 50MHz in the highest frequency range. On power-up, the 50MHz range is by default set, and the instrument can be used without tweaking any settings. The instrument has an extra BNC to be used in the future. A photograph of the V2F instrument is shown in figure 2.

Figure 2. APS Voltage to Frequency Converter.

3.2. *Frequency Noise*

The output signal of the V2F instrument is not a constant frequency. The V2F algorithm does not assure a constant output frequency, but that a certain number of pulses are generated in a given amount of time. Because the output of the V2F is generally connected to digital pulse counter, only the number of pulses in a given time matter in synchrotron experiments. One can think of the V2F as generating an exact frequency plus frequency noise. A figure showing the output signal of the V2F at different frequencies is shown in figure 3.

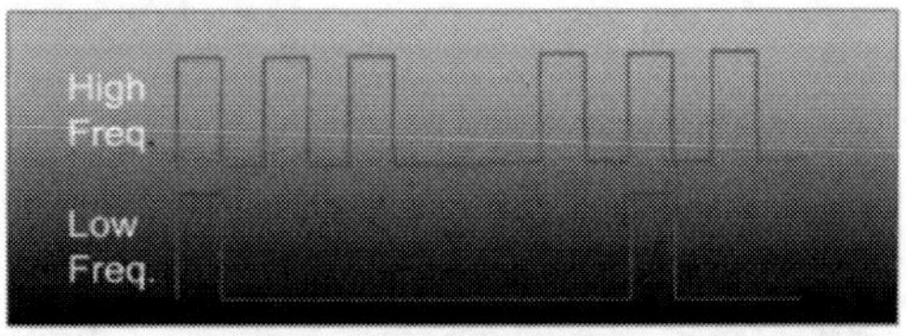

Figure 3. High and low frequency outputs from V2F.

3.3. *Calibration and Specifications*

The instrument is calibrated by supplying a DC voltage of 9V with a voltage standard and adjusting the output frequency to 45MHz. This calibration removes errors in the input amplifier gain. The calibration is stored in the FPGA firmware as a gain factor applied to the digitized voltage. Because connecting two instruments can cause small DC offsets due to amplifier bias and grounding problems, the V2F has the ability to zero the frequency to cancel DC offsets. Because the on-board crystal oscillator can drift with temperature, changing the output frequency, a digital thermometer is included on the instrument, but is currently not used in the calibration. A future firmware revision may include the effects of temperature in the calibration.

The APS instrument can be constructed with either a Linear Technology LTC2400 24-bit converter with a 15Hz sample rate, or a Burr Brown ADS8320 16-bit converter with 100 kHz sample rate, depending on the application [5, 6]. Separate firmware and calibration exists for either converter. The slow 24-bit converter used for slowly varying signals, while the faster 16-bit converter can be used in fly scans, in which a detector signal quickly varies as a sample is rotated on a motor stage.

Table 1. Specifications

Spec	**Value**
ADC Bits	16 or 24
ADC Linearity (24 bit)	30ppm, integral 2ppm, differential
Sample Rate	15Hz, 24bit 100kHz, 16bit
Pulse Output	10ns, TTL
Freq. Ranges	2,10,25,50MHz
Voltage Range	0V-10V

4. Conclusion and Acknowledgements

The V2F Instrument has been in use at Sector 12-ID at APS since 2011. A second instrument resides in the APS detector pool and is currently configured in the faster 16 bit mode. The APS detector group is redesigning the analog portion of the V2F to allow both positive and negative voltages to make the instrument more useful for beam lines. The authors would like to acknowledge Byeongdu Lee and Soenke Seifert for testing the instrument at APS Sector 12. Also, it should be noted that after the APS V2F was designed, a commercial 100MHz V2F instrument was offered by Quantum Detectors [7]. Use of the Advanced Photon Source, an Office of Science User Facility operated for the U.S. Department of Energy (DOE) Office of Science by Argonne National Laboratory, was supported by the U.S. DOE under Contract No. DE-AC02-06CH11357.

5. References

[1] Intruduction to NIM, http://www-esd.fnal.gov/esd/catalog/intro/intronim.htm
[2] Nova R&D corporate website, http://www.novarad.com/
[3] Madden T, Rosenthal D, 1996 U. S. Patent No. 5,512,895
[4] Coln C, Nestler E, 1998 U. S. Patent No. 5,760,717
[5] Linear Technology corporate website, http://www.linear.com/product/LTC2400
[6] Texas Instruments corporate website, http://www.ti.com/product/ads8320
[7] Quantum corporate website, http://www.quantumdetectors.com/

Performance Prediction for a Hockey-Puck Silicon Crystal Monochromator at the Advanced Photon Source

Zunping Liu[1], Gerd Rosenbaum[2], Gary Navrotski[1]

[1] Advanced Photon Source, Argonne National Laboratory, 9700 S. Cass Ave, Lemont, IL 60439, USA

[2] Dept. of Biochemistry, University of Georgia, and Structural Biology Center, Argonne National Laboratory, 9700 S. Cass Ave, Lemont, IL 60439, USA

E-mail: zpliu@anl.gov

Abstract. One of the Key Performance Parameters of the upgrade of the Advanced Photon Source (APS) is the increase of the storage ring current from 100 to 150 mA. In order to anticipate the impact of this increased heat load on the X-ray optics of the beamlines, the APS has implemented a systematic review, by means of finite element analysis and computational fluid dynamics, of the thermal performance of the different types of monochromators installed at the highest-heat-load insertion device beamlines. We present here simulations of the performance of a directly liquid nitrogen-cooled silicon crystal, the hockey-puck design. Calculations of the temperature and slope error at multiple ring currents under multiple operational conditions, including the influence of power, cooling, and diffraction surface thickness are included.

1. Introduction

Cryogenically cooled silicon, high-heat-load monochromators have been successful at 3rd-generation radiation sources [1]. A directly cooled version, the hockey-puck design, has been in use for over ten years at three different beamlines at the Advanced Photon Source (APS) [2]. In general, these cryogenic X-ray optical components take advantage of the simultaneous reduction of thermal expansion coefficient and increase in thermal conductivity for Si [3] while avoiding boiling in the liquid nitrogen (LN_2) coolant. Finite element analysis (FEA) is used to calculate the thermal distortions and resulting surface slope errors of crystals [4-6] under varied operating conditions. As a rule of thumb, for double crystal monochromators, if the calculated slope error is comparable to half of the intrinsic Darwin width for that crystal, a sizable fraction of the flux is lost. In this report we present FEA simulations for the hockey-puck design. Calculations at different ring currents have been carried out for six operational conditions at a macromolecular crystallography beamline, APS Sector 19-ID.

2. The hockey-puck design

The main feature of the hockey-puck design is a flange-like silicon crystal elastically sealed to an Invar base [2], as shown in Figures 1 and 2. The crystal has a relief cut under the diffracting surface creating a thin web and decreasing the amount of power absorbed by the crystal [7]. On the cooling side, the crystal has fins that are sealed into the Invar base, allowing LN_2 to stream past. Sealing is provided by an In-foil-covered, elastic-metal "C-ring" seal clamped between the Si crystal and the

Invar base. This elastic ring compensates for the small thermal expansion mismatch between the parts during cooldown. No direct silicon bonding is required and no LN_2 ports are intrinsic to the crystal.

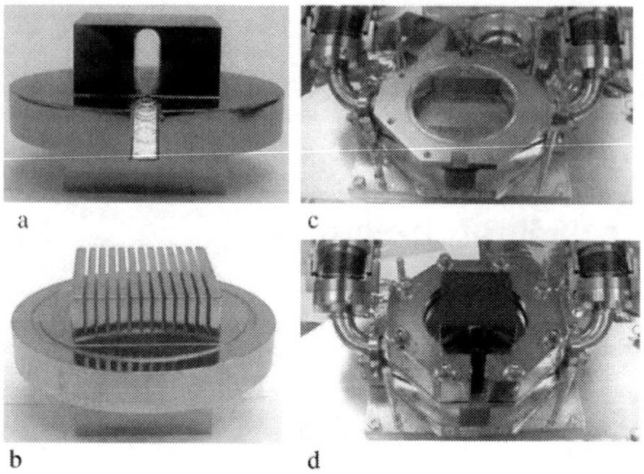

Figure 1. The hockey-puck monochromator. a: crystal with undercut; b: cooling fins; c: cooling system; d: bench assembly (without shield). (As installed at SER-CAT)

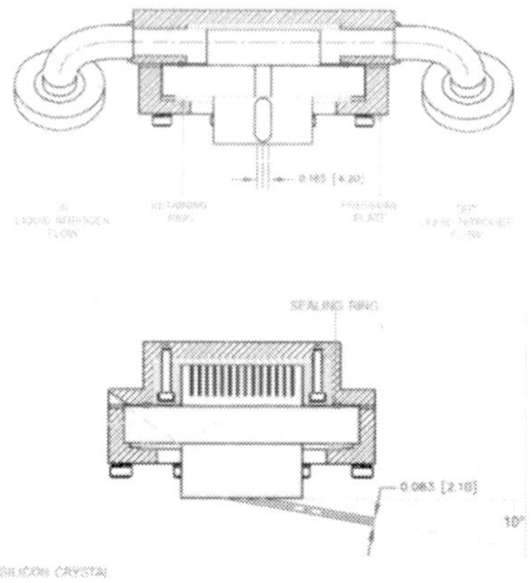

Figure 2. Drawing of the hockey-puck monochromator. The flange-like crystal is mounted to an Invar base with a cut-out pocket.

3. Description of finite element analysis

3.1. Power calculation

FEA calculations start with power generated by APS Undulator A, U33, propagating through a primary aperture of 4.2×2.1 mm² at 50.551 m to the crystal at 52.565 m without intervening windows or filters. Power absorption in the crystal by a layer of arbitrary thickness is calculated over the thermal footprint with SRUFF [8] using the source characteristics and corresponding Bragg angles for six selected energies. Differences in peak power density and total power as calculated by the SRUFF and XOP [9] codes are less than 1% [8]. The calculated power density is fit with a 4th-order Gaussian. Integration over the thermal footprint is the absorbed power by this layer of crystal.

3.2. Heat transfer coefficient

Thermal input parameters for FEA analysis are calculated conventionally using i) the known crystal geometry, ii) experimentally set liquid nitrogen coolant conditions (volume, velocity, temperature, pressure), and iii) derived values of hydraulic diameter, Reynolds number, and Nusselt number, to yield a calculated heat transfer coefficient, h=0.0043 W/(mm² K). In this paper, a more conservative heat transfer coefficient of h_{eff} = 0.003 W/(mm² K) is used for the majority of the calculations.

3.3. Model for finite element analysis

FEA analysis includes two steps: steady state thermal analysis followed by static structural analysis. For power absorption, a volumetric absorption model has been found to be better than either surface absorption or staggered layer models [5]. For absorption by a 1-mm-thick web, the absorption profile is split into ten 0.1-mm-thick layers, and the first layer is further split into ten 0.01-mm-thick layers, as shown in Figure 3. For the static structural analysis, zero vertical displacement is assigned at the location of the sealing ring and zero radial displacement at the geometric center of the sealing circle.

4. FEA predictions and discussion

A traditional FEA approach is used here, calculating surface temperatures, thermal strains, vertical displacements, and thermal slope errors along the meridian axis of the beam footprint.

Figure 3. FEA model of the hockey-puck Si crystal with an applied heat load. Left: Fine mesh in the footprint region and coarse mesh elsewhere. Middle: Magnified view of the heat loading for volumetric absorption. The first ten layers are each 0.01 mm thick, and the next nine layers are each about 0.1 mm thick. Right: power density scale corresponding to a total absorbed power of 132 W from the undulator with 3rd harmonic at 30 keV and 100-mA ring current.

4.1. Initial cooldown

A deformation calculation during cooldown from zero strain at room temperature to thermally strained at cryogenic temperature is –0.000231 from the FEA model and –0.000227 from analytical calculation; a reasonable agreement. The resulting slope change of the first crystal surface, due to the geometry of the system and cooldown deformation, is expected to be zero as confirmed by FEA.

4.2. Power effect

The slope error vs. power curve can generally be divided into three regions: linear, transition, and nonlinear [5, 6]. In the linear region, slope error increases with power. As power continues to increase, the slope error decreases in the transition region because the silicon expansion coefficient changes from negative to positive. Here, the thermal contraction (negative expansion) dominates the thermal deformation. In the third (nonlinear) region, the slope error increases rapidly with power as the contraction no longer dominates the thermal deformation. Since the slope error vs. power curve varies with the footprint size [5, 6], a modified linear power density (MLPD) was proposed to replace the total absorbed power [6]. The calculated peak-valley thermal slope error (Table 1) versus MLPD for the hockey-puck design is shown in Figure 4. It also generally falls into three categories: linear (< 35 W/mm), transition (~35 - 53 W/mm), and nonlinear (>53 W/mm). Increasing the cooling efficiency, h_{eff}, changes these transitions (Figure 5). If the maximum value of slope error from the linear region is treated as one of failure criteria for the crystal, the hockey-puck crystal will work well at 150 mA. The thermal slope error is less than half of the Darwin width for a majority of the cases tested. It is known that the slope error is independent of the cooling coefficient in the linear region but not in the nonlinear region [5, 6]. For example, the slope errors at 13.474 keV in Table 1 show a strong nonlinear relationship when the storage ring current increases from 100 mA to 200 mA.

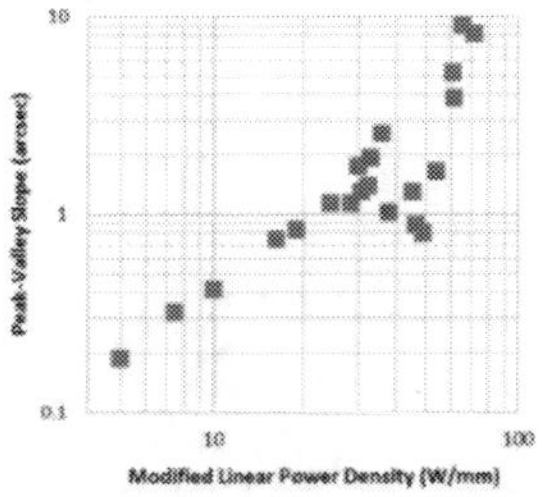

Figure 4. Thermal slope error vs. MLPD (includes all Table 1 cases).

5. Summary

The hockey-puck crystal works well at 150 mA in terms of temperature and thermal slope error. The results in Table 1 predict that the temperature at the cooling fins will remain less than 85 K, the boiling point of LN_2 at 27 psi, for all energy selections at storage ring currents up to 200 mA. The thin-web undercut aids in maintaining this low interface temperature by lowering absorption and by elongating the thermal conduction path. Increasing web thickness causes a corresponding increase in temperature at the convection interface and an expected decrease of peak temperature and slope error at the diffraction surface. If the maximum value of slope error from the linear region is treated as an acceptable value for the crystal, the hockey-puck crystal works well for all energy selections at 150 mA and for 12.66 keV up to 200 mA.

Table 1. Effect of power on the hockey-puck crystal with a 1-mm-thick web.

Energy (keV)	Current (mA)	Peak power density* (W/mm²)	Power (W)	MLPD (W/mm)	Peak temperature (K)	Lowest temperature (K)	Slope error (arcsec)	Darwin width (arcsec)
6.5 (1st harm.)	100	1.52	216	35.84	114.11	80.364	2.55	8.52 (Si-111)
	150	2.28	324	53.76	144.14	81.68	1.65	
	200	3.04	432	71.68	195.54	82.985	8.23	
12.66 (1st harm.)	100	0.24	48	4.99	82.21	78.299	0.19	4.22 (Si-111)
	150	0.36	72	7.48	84.992	78.595	0.32	
	200	0.48	96	9.97	87.656	78.893	0.42	
13.47 (3rd harm.)	100	0.99	304	30.15	113.69	81.456	1.74	3.96 (Si-111)
	150	1.49	456	45.23	141.58	83.278	1.29	
	200	1.98	613	60.75	192.21	85.119	5.25	
19.50 (3rd harm.)	100	0.87	253	18.99	99.967	80.813	0.83	2.72 (Si-111)
	150	1.31	380	28.49	115.05	82.348	1.14	
	200	1.75	507	37.99	134.12	83.865	1.02	
30.0 (3rd harm.)	100	0.72	132	16.24	93.542	79.334	0.75	0.37 (Si-333)
	150	1.07	198	24.36	104.2	80.145	1.14	
	200	1.43	264	32.48	116.13	80.952	1.39	
30.0 (5th harm.)	100	1.30	250	30.80	111.55	80.78	1.29	0.37 (Si-333)
	150	1.94	376	46.21	138.2	82.299	0.89	
	200	2.59	501	61.61	182.13	83.801	3.86	

*: The peak power density is for absorption by the first layer of 10 μm.

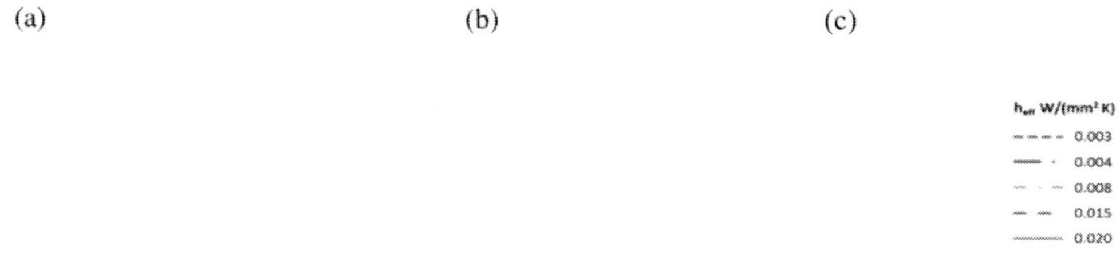

Figure 5. Cooling value effect, h_{eff}, on (a) temperature, (b) thermal strain and (c) slope error (Condition: energy 13.474keV, 3rd undulator harmonics, Si-111, current 100mA).

Acknowledgments
The authors thank Mati Meron for utilizing his code SRUFF and Catherine Eyberger for edit. The Advanced Photon Source and the Structural Biology Center at Sector 19 are Office of Science User Facilities operated for the U.S. Department of Energy, Office of Science and Office of Biological and Environmental Research by Argonne National Laboratory under Contract No. DE-AC02-06CH11357.

References

[1] Bilderback D H *et al* 2000 *J. Synchrotron Rad.* 7 53
[2] Ivanov I *et al* 2000 *AIP Conf. Proc.* **521** 271; Rosenbaum G *et al* 2006 *J. Synchrotron Rad.* **13** 30
[3] Bilderback D H 1986 *Nucl. Instrum. Methods* **246** 434
[4] Tajiri G *et al* 2001 *J. Synchrotron Rad.* **8** 1140
[5] Zhang L *et al* 2003 *J. Synchrotron Rad.* **10** 313
[6] Huang R and Bilderback D H 2012 *Proc. of SPIE* **8502** 85020B
[7] Knapp G S *et al* 1994 *Rev. Sci. Instrum.* **65** 2792
[8] Jaski Y *et al* 2002 *Proc. of MEDSI 2002* 2002 (Argonne National Laboratory, ANL-03/7) p 390
[9] Sanchez del Rio M and Dejus R J 1997 *Proc. of SPIE* **3152** 148

Journal of Physics: Conference Series **493** (2014) 012010 doi:10.1088/1742-6596/493/1/012010

Variable q-range x-ray scattering chamber *for chemical and materials science at the Advanced Photon Source*

O Schmidt[1], J DeBartolo[2], C Kurtz[2], B Lee[2], S Seifert[2], R Winans[2] and X Zuo[2]
[1]Engineering Support Division, Argonne National Laboratory, Argonne, IL, USA
[2]X-Ray Science Division, Argonne National Laboratory, Argonne, IL, USA

E-mail: oschmidt@aps.anl.gov

Abstract. We present here the design of a novel variable q-range x-ray scattering instrument recently installed at Sector 12 of the Advanced Photon Source. This device provides automated, computer-controlled q-range changes for x-ray scattering experiments by varying the sample-to-detector distance within a large vacuum chamber. Eliminating the need to vent the system when changing camera lengths allows for quick and efficient change-overs between experimental setups. The detector cannot operate in a vacuum environment; therefore it is housed within an air chamber open to atmospheric pressure. A large carbon window isolates the detector from vacuum while allowing high x-ray transmission. An array of motorized beam stops mounted directly upstream of the window protects the detector from the direct x-ray beam for various types of scattering experiments. A smaller detector protrudes into the lower front section for simultaneous wide-angle x-ray scattering data collection. A fully automated support structure aligns the vacuum chamber to the x-ray trajectory.

1. Introduction

X-ray scattering has proven to be a premier measurement technique for quickly obtaining information about the structure and composition of materials. In an x-ray scattering experiment, a sample is illuminated by a beam of x-ray photons, and the scattering pattern that is produced is measured as a function of the angle between the incident and scattered beams. The length scale that is examined within a sample is a function of the photon energy and the scattering angle; thus, studying complex materials in a comprehensive way often requires a combination of small and wide-angle measurements.

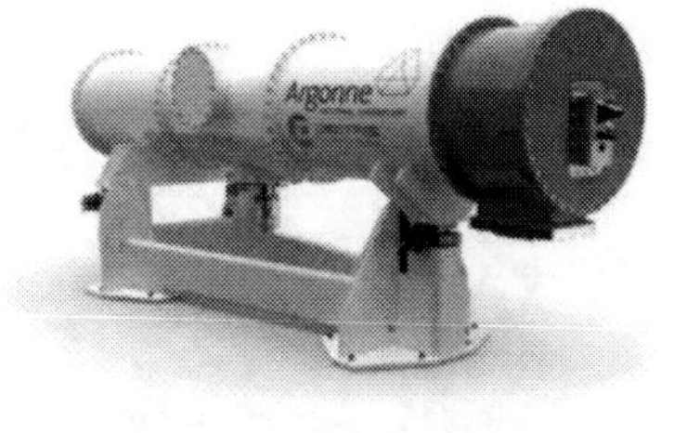

Figure 1. X-ray Scattering Chamber

Small-angle x-ray scattering (SAXS) is useful for obtaining nanoscale structural information for a variety of samples, while wide-angle x-ray scattering (WAXS) resolves features in the sub-nanometer regime. An area detector collects data that spans a particular angular range; this range depends on the size of the detector and its distance from the sample. The angles that are measured by the detector are proportional to the scattering vector, q. With fixed energy, the q-range can be manipulated by changing the distance between the sample and the detector; this allows for different sizes of interest within a sample to be probed. An evacuated flight path is required downstream of the sample to reduce background scattering from air and to allow for detection of a clear signal from the sample.

Sector 12 at the Advanced Photon Source at Argonne National Laboratory was recently upgraded to include a new, fully dedicated x-ray scattering station. Completed in the fall of 2010, beamline 12-ID-B was formed from canting beamline 12-ID. The beamline utilizes a large offset side-bounce monochromator and a set of focusing mirrors [1] to deliver desired beam to the endstation. The available energy range at this beamline is 7.8 to 14 keV, but the beamline is optimized for and typically run at a fixed energy of 12 keV. At this energy range, a vacuum chamber is necessary to prevent scattering and absorption by air.

The diversity of themes being studied at 12-ID-B together with the high demand for beam time requires a setup that incorporates quick and simple sample-to-detector distance changes and efficient changeovers between experimental setups.

2. Construction

Similar in design to the much larger neutron scattering chambers at the National Institute of Standards and Technology [2], the variable q-range x-ray scattering chamber, shown in Figure 2, was built to facilitate rapid sample-to-detector changes by housing the detector inside the vacuum chamber, so that the q-range can be varied without the need to vent the system. This feature is unique to the Sector 12 instrument and enhances the efficiency and user experience.

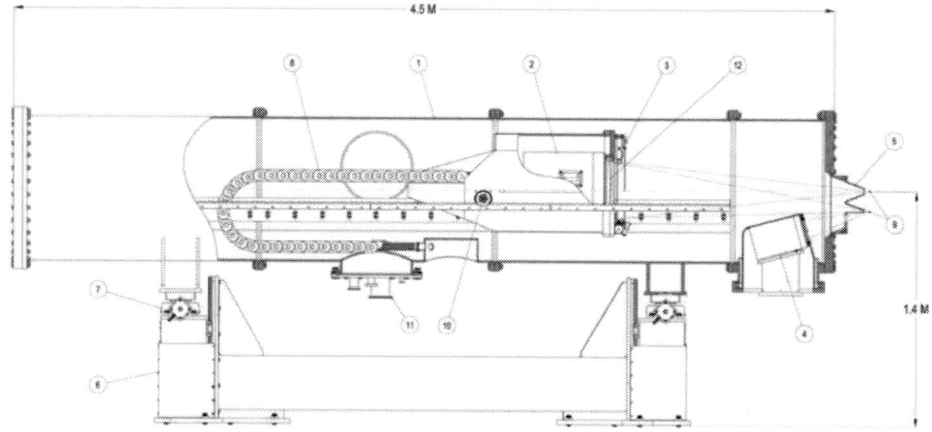

Figure 2. Variable q-range x-ray scattering chamber cutaway: 1) vacuum chamber, 2) SAXS detector housing, 3) beamstops, 4) WAXS detector housing, 5) interface flange, 6) support structure, 7) kinematic mount, 8) flex hose/cable carrier, 9) sample position, 10) gear drive and rack, 11) feedthrough ports, 12) carbon window.

2.1. *Vacuum chamber*

The vacuum chamber provides a scatter-free flight path between the experimental sample and the detector. The camera length of the main SAXS detector can be continuously varied from 400 mm to 3.5 meters along the length of the chamber. To minimize overall weight, ease fabrication, and provide significant cost savings, the chamber was fabricated in four main sections out of 6061 aluminum. O-rings provide all of the vacuum seals, and various ports allow for electrical feedthroughs, vacuum connections, and maintenance access. A rectangular flange on the bottom of the upstream section allows for a smaller removable detector housing for simultaneous WAXS data collection. Average pump down to 10^{-4} Torr is commonly achieved within 24 hours.

2.2. *SAXS detector housing*

Not designed for in-vacuum applications, the Pilatus 2M detector is nestled within a welded aluminum housing maintained at atmospheric pressure. A large 13-in-diameter carbon window, comprised of 3 layers of 125-μm-thick unidirectional carbon-fiber sheet, isolates the detector from vacuum and minimizes x-ray absorption. Data cables, power, and water cooling lines enter the main vacuum chamber from underneath and are fed to the detector housing through three flexible stainless steel vacuum hoses. The hoses are mounted within a flexible cable guide to prevent kinking and to allow for the full range of detector travel along the length of the vacuum chamber. The detector housing is supported and guided by a single linear rail on one side, eliminating the need to precisely align two rails over the large distance. The other side of the box rests on a NEXEN® precision roller pinion system that moves the chamber via direct drive from an in-vacuum stepper motor. An inline 50:1 gearbox provides a linear resolution of 12 μm/step, and the position can be precisely monitored by means of a rotary encoder fixed to the rear output shaft of the motor. The rail and rack are aligned to within 200 μm in both lateral and elevation, allowing q-range adjustment of the detector along its entire travel range without the need to realign the beam stops.

2.3. Beamstops
Remotely operated beamstops mounted directly upstream of the carbon window protect the detector from the direct x-ray beam. The beamstops are mounted to motorized mechanical slides by thin stainless steel rods. Each beamstop has a minimum travel of ± 20 mm and can be driven completely out of the detector area. A retractable aluminum plate shields the detector during beamstop alignment. For grazing incident measurements, a long 3-mm-wide tungsten strip, coupled to the horizontal and vertical motion of the lower beamstop, can be rotated upwards to block the reflected beam.

2.4. WAXS detector housing
To simultaneously collect wide-angle scattering data, the smaller Pilatus 300K detector protrudes into the lower front section of the main vacuum chamber. The detector is positioned to minimize dead pixel space between SAXS and WAXS measurements. A smaller carbon window isolates the detector from vacuum. The housing is removable to allow the main detector to achieve shorter distances to the sample if required.

2.5. Interface flange
Interchangeable inlet cones mounted to the upstream flange of the vacuum chamber accommodate various scattering angles and sample distances while minimizing the air gap between the sample and the detector. Kapton, mica, and/or Mylar windows glued to the apertures provide an adequate vacuum seal and minimize x-ray absorption. A thin tungsten septum plate installed within the cone blocks the possible scattering produced from the direct beam on the window to the WAXS detector.

2.6. Support structure
The support structure is designed to provide vertical as well as pitch, roll, and yaw manipulation of the chamber in order to align the detector travel axis to the x-ray beam trajectory. Three independently operated lifting columns are mounted to a rigid, welded steel frame. Each lifting column consists of a carrier, guided by a pair of vertically mounted linear rails, and driven by a 24:1 machine screw jack directly coupled to a high torque Nema 34 size stepper motor. Each column has an overall travel range of 12 inches, a load capacity of 1500 lbs, and a vertical resolution of 1.3 μm per motor half step. Both the aluminum vacuum chamber and the steel support base are externally powder coated to match for a consistent finish across the device.

2.7. Kinematic mounts
Arranged in the typical cone, vee, and flat configuration and affixed to the top of each lifting column, special kinematic mounts were designed in order to fully constrain the system, yet allow for the full range of motion required. Crossed roller bearing slides, mounted to efficiently designed aluminum frames, allow for a maximum travel of ± 1.5 inches in the x and z directions, and a maximum load capacity of 1000 lbs. each. A large spherical bearing allows for a tip-tilt misalignment of up to 8° and

an in-plane rotational misalignment of 15°. An integrated spherical rod end, aligned at the center of the spherical bearing, captures the assembly in the unlikely chance that the chamber would lift off. Over-travel is prevented by limit switches mounted to each axis, backed up by internal hard-stops. The cone and vee supports, driven via direct-drive hybrid linear stepper actuators, provide additional yaw alignment capability, with a resolution of 12.7 μm per step. As the vertical columns are independently actuated, the kinematic mounts fully constrain the chamber while allowing for the necessary degrees of freedom.

3. Operation

The beamstops, nozzles, and support structure motions allow the SAXS chamber to operate in three experimental modes, providing measurement capabilities for centered SAXS, off-centered SAXS/WAXS, and GISAXS. When the setup is at its centered configuration, the x-ray beam enters the chamber through the top nozzle, and the detector is protected by a small 3-mm-diameter beamstop located just upstream of the center of the detector face. Using this configuration, data is collected over the entire range of the detector; as such, the images contain information for the full radius of scattering, which is often required for anisotropic samples.

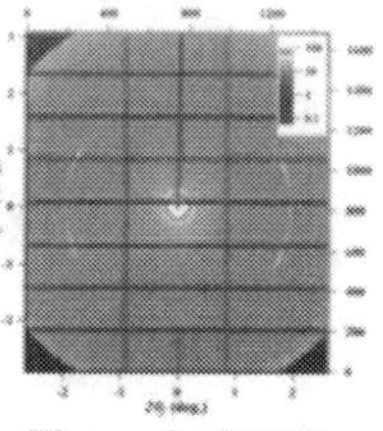

Figure 3. SAXS

To run off-centered SAXS/WAXS, the chamber is raised so that the beam enters the lower nozzle and window. A corresponding 3-mm-diameter beamstop is positioned to block the direct beam. This offset configuration provides a wide dynamic q-range, as the lower beam position allows for a greater scattering angle to be collected at the SAXS detector. Furthermore, this setup makes use of the WAXS detector, which is installed in the lower front section of the chamber. Merging data from both detectors provides a q_{max}/q_{min} ratio of about 150 and an overall q-range from 0.003 to 2.4 Å^{-1}.

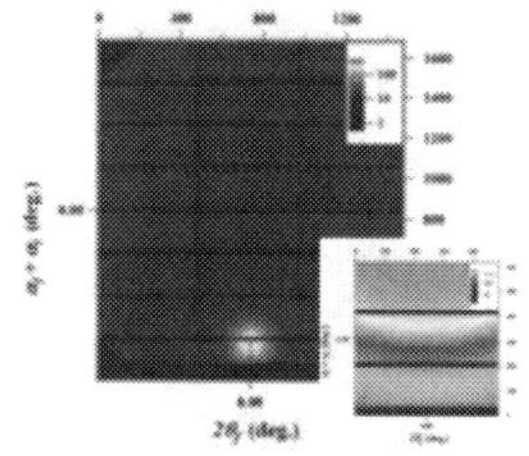

Figure 4. SAXS/WAXS

In addition to transmission SAXS/WAXS, the chamber is also equipped for collecting data in grazing incidence geometry. When the beamline runs GISAXS, an additional rotary motor is used at the sample area for careful alignment of films and flat surfaces. The scattered beam enters the chamber through the lower nozzle, and a long beamstop strip is used concomitantly with the off-centered SAXS beamstop to block the reflected beam. GISAXS measurements are useful for characterizing density correlations and analyzing nanoscale features at the surface of a sample or at an interface within a sample.

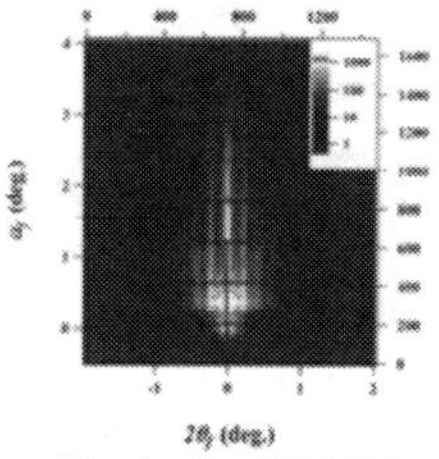

Figure 5. GISAXS

4. Conclusions

The installation of the SAXS chamber has greatly improved the versatility and efficiency of the 12-ID-B beamline. Because the setup can be changed quickly and accurately, a wider array and larger number of user groups can be easily accommodated. Additionally, many users take advantage of the chamber's capability for simultaneous SAXS/WAXS measurements. The quality of the data has also improved, due to the accuracy by which alignment and scanning procedures can be done.

5. References

[1] Trakhtenberg E, Collins J, Schmidt O 2010 A Novel Precision Bender System for Focusing Mirrors *Proc. of MEDSI-2010,* Oxford, United Kingdom doi:10.1017/S2044820110000262

[2] Glinka C J, Barker J G, Hammouda B, Krueger S, Moyer J J and Orts W J 1998 The 30 m Small-Angle Neutron Scattering Instruments at the National Institute of Standards and Technology *J. Appl. Cryst.* **31** 430 doi:10.1107/S0021889897017020

17th Pan-American Synchrotron Radiation Instrumentation Conference (SRI2013) IOP Publishing
Journal of Physics: Conference Series **493** (2014) 012035 doi:10.1088/1742-6596/493/1/012035

Expected performance of cryogenic silicon monochromator for high power density and high coherence ERL beamlines

Rong Huang[1], **Kenneth Finkelstein**[1] **and Donald H. Bilderback**[1,2]

[1]Cornell High Energy Synchrotron Source, Cornell University, Ithaca, NY14853, USA
[2]School of Applied Engineering Physics, Cornell University, Ithaca, NY 14853, USA

Email: rh66@cornell.edu

Abstract. With small emittance and a high degree of X-ray coherence, the Energy Recovery Linac beamlines, proposed at Cornell, need excellent optics to preserve the high quality of X-ray beam. In this paper, thermal deformation of cryogenic silicon monochromators and its effect on wavefront deformation are studied. For crystal thermal deformation analysis, we introduce the so-called modified linear power density as a universal parameter of heat load. For the wavefront simulation, using synchrotron radiation workshop, we study the focusing of source under one-to-one demagnification ratio, for the full central cone of undulator radiations. The studies conclude that cryogenic Si monochromators, in general, are suitable for Cornell energy recovery linac beamlines.

1. Introduction

Future Energy Recovery Linac (ERL) beamlines, with smaller source emittance and higher degree of X-ray coherence than available at existing 3rd generation sources [1], need excellent optics to preserve the high quality of the X-ray beam. For Cornell ERL, 5m and 25m long undulators with 19 mm periods will produce extremely low-divergent central cones with very high power density; therefore, the potential thermal deformation of monochromator crystals has to be carefully studied. LN_2-cooled Si crystal monochromators have been widely used at 3rd generation sources with success [2,3], and the perfect atomic structure of Si crystals is good for preserving X-ray wavefront. Since crystal thermal deformation is sensitively related to many factors, including power density, total power, and the dimensions of heating footprint, we introduce the so-called modified linear power density (MLPD), with which the Si thermal deformation can be better understood with given heat load.

With crystal deformation profiles calculated with ANSYS, the wavefront distortion can be evaluated with the simulation tool, Synchrotron Radiation Workshop (SRW) [4]. In this paper, the wavefront distortion by monochromator is evaluated by simulating a one-to-one focusing of the X-ray source through an ideal thin lens set right after monochromator.

2. Modified linear power density and the thermal deformation of cryogenic Si monochromator

Cornell ERL is proposed to run with current of either 25mA (high coherence mode) or 100mA (high flux mode). ERL round source allows a unique Delta undulator design with its polarization changeable between planar mode and helical mode [1]. For clarity, major source and undulator parameters are summarized in table 1

The on-axis radiation power density of CHESS ERL undulators can be as high as 1.1×10^6 W/mrad^2 for 25m undulator at planar mode and 3.4×10^5 W/mrad^2 at helical mode. As a comparison, APS undulator A maximal on-axis power density is about 1.6×10^5W/mrad^2. On the other hand, because of the small ERL emittance in both transverse directions which creates small radiation cone, the maximal total power in central cone is about 400 W and 95W for ERL planar and helical undulators respectively, comparable to the heat load on third generation sources. There will be 8 different kinds of beamlines for Cornell ERL out of the different combinations of machine currents, undulator lengths and undulator polarization modes. Before we start to use ANSYS to simulate all the 8 kinds of beamlines, we use a simplified model to understand what we can expect for cryogenic monochromators. With the assumption that the surface heating area is much smaller than crystal dimensions, the maximum temperature at the center of thermal footprint can be estimated as [5]:

$$T = T_k + (T_0 - T_k)\exp(0.0183 P_M), \quad (1)$$

with

$$P_M = [1 + ln(L/W)/2.571]Q/L, \quad (2)$$

which we call modified linear power density (MLPD). Q is the heating power; *L* and *W* are the length and width ($L \geq W$) of X-ray footprint, and T_0 is the temperature at the "far-away" crystal boundary and T_k is a constant of 47.4K. The coefficient in eq.(1) is calculated with P_M in the unit of W/mm. Eq.(1) tells us that it is the linear power density that determines crystal temperature increase.

Table 1. Cornell ERL source and undulator parameters

Electron source parameters							
Mode	$\Delta E/E$ (%)	Current(mA)	$\varepsilon_x/\varepsilon_y$ (nm rad)	σ_x(μm)(S/L)[a]	σ_y (μm)(S/L)[a]	σ'_x(μrad)(S/L)[a]	σ'_y (μrad)(S/L)[a]
High flux	0.0186	100	0.031/0.025	4.95/11.1	4.45/9.95	6.22/2.78	5.59/2.50
High coh.	0.0088	25	0.013/0.011	3.217/7.19	2.904/6.49	4.042/1.81	3.649/1.63

Undulator parameters						
Type	Length (m)	Period (mm)	Period number	Gap (mm)	B_{max}(T) (P/H)[b]	K_{max} (P/H)[b]
Short	5	19	263	5	1.2524/0.8856	2.222/1.571
Long	25	19	1315	5	1.2524/0.8856	2.222/1.571

[a]Source dimensions are calculated with β functions set as undulator length over 2π, and "S/L" stands for short/long undulators; [b]"P/H" stands for planar/helical undulators.

Even eq.(1) is only accurate within a limited range of heat load [5], we found out that the MLPD, P_M, can be used to parameterize crystal thermal deformation when the assumption of surface heating is reasonable. The power load for ERL helical undulators, and for planar undulators not working at high value of deflection parameter K, can be approximated as surface heat. With MLPD, the severity of crystal thermal deformation can be evaluated independent of facilities and beamlines. It is believed that cryogenic Si crystal thermal deformation can be divided into three regions [6]. When heat load is small, the slope errors caused by thermal deformation increase linearly with heating power. As power increases to a level called transition region, the slope errors may not increase and could even decrease as heating power increases. Further increase the power to the so-called non-linear region, or run-away region, the slope errors increase dramatically as heating power increase. In terms of MLPD, we find out that the crystal deformation is in the linear region when MLPD is around 50 W/mm or less, and in the transition region when MLPD is around 50 – 100 W/mm and in the nonlinear region when MLPD is around 100 w/mm or higher [5].

The results of ANSYS simulations for CHESS ERL beamlines are summarized in table 2, with the I-weighted slope errors defined as $\alpha_{RMS} = \sqrt{\int I(x)[E(x) - mean(E(x))]^2 dx \Big/ \int I(x) dx}$, where *I(x)* is the monochromatic X-ray intensity, and *E(x)* is surface slope, along position x on a crystal surface.

It can be seen from table 2 that the MLPD will be larger than 100 W/mm only when ERL machine runs at 100 mA, and meanwhile the planar undulators work at maximum K. Under that circumstance, the cryogenic mono will work in the nonlinear region with large deformation. If any future ERL beamlines need to have cryogenic Si monochromator work at this extreme condition, possible ways to mitigate may include positioning monochromator further downstream of the beamline to reduce MLPD, using internal cooled Si crystal to increase cooling efficiency, and reducing the openings of power slit.

Table 2. The summary of ERL monochromator slope errors simulated using ANSYS

Undulator length (m)	Current (mA)	Mono to source (m)	Total power (W)	MLPD (W/mm)	I-weighted slope errors (μrad)	1st harmonic energy (keV)
Helical undulator, K = 0.707, maximum on-axis power density condition						
25	25	45	24	12	0.52	8.33
25	100	45	95	47	2.05	
5	25	35	24	6.7	0.30	
5	100	35	95	27	1.25	
Planar undulator, K=2.22, "closed-gap" condition						
25	25	45	98	78	1.35	3.6
25	100	45	392	310	80	
5	25	35	99	45	0.9	
5	100	35	395	178	12.5	
Planar undulator, K=1.06.						
25	25	45	35	17	0.61	8.0
25	100	45	141	70	1.94	
5	25	35	35	10	0.35	
5	100	35	140	40	1.47	

3. Wavefront propagation through cryogenic monochromator

ERL ultra-low emittance is advantageous to the needs of high degree of transverse coherence and high-flux nano-focusing, which in turn requires high quality optics to maintain X-ray beam quality. In general, the Si crystal thermal deformation can be considered as a combination of a cylindrical curvature and a residual figure after cylindrical curvature subtraction. While the wavefront change by cylindrical curvature can be corrected with downstream focusing or defocusing optics, the changes arising from the residual figure may not be easily correctable, with a net effect of increase beam size in focusing. Using SRW [4] we simulate X-ray propagation starting from undulator, through monochromator crystal and a "perfect thin lens", to make a 1:1 focusing of the source. We then evaluate the image changes caused by monochromator deformations.

Fig.2 shows the wavefront simulation results of 1:1 focusing for CHESS ERL 5 m long helical undulator source, working at the highest possible thermal load condition with K=0.707. For comparison, results with considering wavefront deformation from monochromator crystal and without considering the deformation are presented together in fig.2. It can be seen that when ERL machine runs at high coherence mode with current of 25 mA, there are almost no changes of focusing after considering wavefront deformation from monochromator crystal. For ERL machine running at 100 mA high flux mode, there are only minor changes after considering crystal deformation, with an increase of FWHMx from 15.4 μm to 15.6 μm, and FWHMy from 14.2 μm to 14.7 μm.

Similar wavefront simulations have been done for 25m long helical undulators and also for planar undulators with 1st harmonic X-ray energy at 8 keV. The results are summarized in table 3, from which it can be seen that when ERL machine runs at 25 mA, the crystal thermal deformation has negligible effect on wavefront deformation. Minor increase of focal size can be seen when ERL machine runs at 100 mA high flux mode.

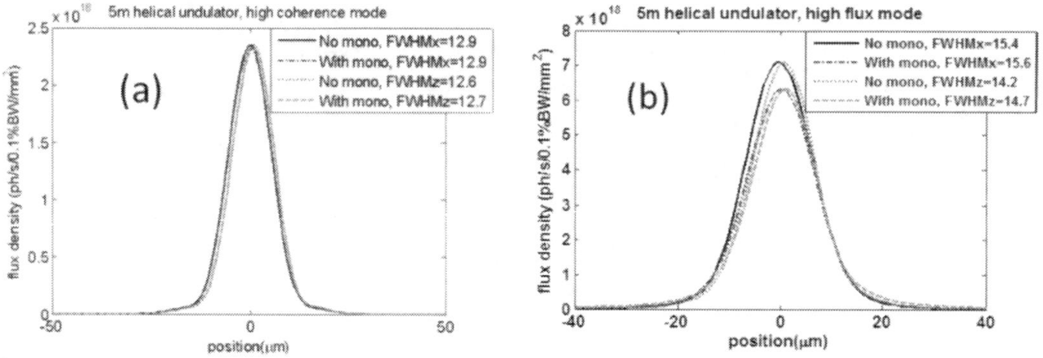

Figure 1. Wavefront simulations of Cornell ERL 5m long helical undulators at 8.32 keV. (a) The comparison of the 1:1 focusing of the source with and without considering Si crystal thermal deformation, when ERL machine runs at 25 mA high coherent mode. (b) The same simulation for ERL machine working at 100 mA high flux mode.

Table 3. Summary of wavefront simulation with and without considering crystal deformations

Undulator type	Current (mA)	1st harmonic energy (keV)	Without crystal deformation FWHM$_x$(μm)	Without crystal deformation FWHM$_z$(μm)	With crystal deformation FWHM$_x$(μm)	With crystal deformation FWHM$_z$(μm)
5m helical	25	8.32	13.1	12.9	13.2	12.9
	100		15.4	14.2	15.6	14.7
25 helical	25		28.8	27.9	29.2	27.8
	100		34.7	32.1	34.8	33.2
5m planar	25	8.0	12.9	12.6	12.9	12.7
	100		15.4	14.4	15.9	15.2
25m planar	25		28.4	27.5	28.7	27.4
	100		35.0	32.4	35.1	34.0

4. Summary

Cryogenic Si monochromators in general will be suitable for CHESS ERL beamlines with small slope errors and negligible or minor wavefront distortions.

Acknowledgments

Rong Huang acknowledges the help from Dr. Oleg Chubar for the use of SRW. CHESS is supported by the NSF & NIH/NIGMS via NSF award DMR-0936384 and NSF DMR-0807731.

References

[1] Bilderback DH, Brock JD, Dale DS, Finkelstein KD, Pfeifer MA and Gruner SM 2010 *New J Phys* **12** 035011
[2] Chumakov A et al 2004 *J. Synchrotron Rad.* **11** 132
[3] Tamasaku K et al 2002 *Proc. of SPIE* **4782** 132
[4] Chubar O *et al* 2002 *Proc. of SPIE* **4769** 145
[5] Huang R and Bilderback DH 2003 *Proc. of SPIE* **8502** 8502B-1
[6] Zhang L, Lee WK, Wulff M and Eybert L 2003 *J. Synchrotron Rad.* **10** 313

Experience with the CSPAD during dedicated detector runs at LCLS

G A Carini, S Boutet, M Chollet, A Dragone, G Haller, P A Hart, S C Herrmann, C J Kenney, J Koglin, M Messerschmidt, S Nelson, J Pines, A Robert, S Song, J B Thayer, G J Williams, D Zhu

SLAC National Accelerator Laboratory, 2575 Sand Hill Rd, Menlo Park, 94025 CA

E-mail: carini@slac.stanford.edu

Abstract. In-house developed cameras and other commercial detectors are typically tested with x-ray tubes and at synchrotron beamlines before being deployed and used for science experiments. In a prototyping phase, this is needed to understand and characterize the behavior of the detector. In a more advanced development phase, measurements with x-rays are required to characterize and calibrate the camera. Tests at synchrotron beamlines in actual experimental conditions are indeed a valuable source for detector developers. However, when all photons arrive at once, as for FELs, the response of the detector can be very different from that obtained with a synchrotron beam which behaves more like a CW (continuous) source. This behavior was already observed during users runs at LCLS and recently investigated during dedicated detector beamtime. The linearity of the response of the Cornell-SLAC Pixel Array Detector (CSPAD) was investigated. Results are presented and discussed.

1. Introduction

With the advent of x-ray Free Electron Lasers (FELs), it was clear that a dedicated class of detectors had to be developed to fully exploit the source characteristics. The pulsed nature of these sources requires the use of purely integrating x-ray detectors. Depending on the experiment, a suite of detectors with imaging capability has to cover applications characterized by low signal or very high maximum signal, typically with a good spatial resolution. Detecting intense signals in short times poses some new challenges for the silicon diodes, the frontend electronics and the overall camera concept. The response of the detector under these operating conditions can be simulated; although due also to the novelty of the source, comprehensive predictions are difficult. The consequences of operating at these conditions need to be understood and validated by experimental investigation with x-rays so that new generation of detectors can by optimally designed. For this reason tests of cameras developed for FELs with x-ray laboratory sources and at synchrotron beamlines are not sufficient for a complete characterization of these detectors; and the final validation of the detector concept is often achieved during the experiments [1-3]. At the Linac Coherent Light Source (LCLS) the great majority of hard x-ray experiments employs a variant of the Cornell-SLAC Pixel Array Detector (CSPAD) [4-9]. The characteristics of this camera (currently available in three sizes: 2.3 Mpixel, 560 kpixel and 140 kpixel) are summarized in Table 1. Figure 1 shows the schematic of a pixel.
In order to analyse and characterize the behaviour of the CSPAD, a series of measurements were carried out at the Stanford Synchrotron Radiation Lightsource (SSRL) and at LCLS with particular

attention to the case of high photon fluences where differences can be expected. Details of the experimental setups and test results are presented and discussed in the following paragraphs.

Table 1. CSPAD properties

Pixel size	110 μm x 110 μm
Area	326 cm^2 (2.3 Mpixel)
Maximum signal	2700 8 keV photons/pixel (low gain) 350 8 keV photons/pixel (high gain)
Frame rate	120 Hz
Noise	~ 3.5 keV (low gain), ~ 1 keV (high gain)

2. Experimental evaluation with x-rays

2.1. Stanford Synchrotron Light Source

A series of tests were performed on CSPADs at SSRL beamlines BL 2-2 and 7-2. Typical working conditions at synchrotrons offer a relatively stable beam, whose intensity is easily measured with ion chambers. Detectors quantum efficiency at different energies can be evaluated with fast scans by simply rotating a monochromator crystal. Figure 2 shows the normalized response of the CSPAD-140k v1.5 (a version of the CSPAD with upgraded electronics, ASIC and firmware) at energies ranging from 7 to 15 keV. The measurements were performed at room temperature. The figure shows histograms of all pixels with pedestal and common mode correction without any per-pixel correction. It can be noted that although the detector is not designed to tackle spectroscopy applications all energy lines are well resolved. To study the linearity of the response of CSPADs scans sweeping the integration time were performed. These measurements allowed improving a previous version of this detector, which had shown non-linearity at low intensities. Results from scans in high and low gain of a CSPAD-140k v1.2 (a version of the CSPAD with upgraded electronics and firmware) are shown in figure 3. They both present saturation at 14000 ADU. Thanks to the short integration time, the attenuated direct beam was used for all the measurements. Harmonic contamination in the beam was negligible.

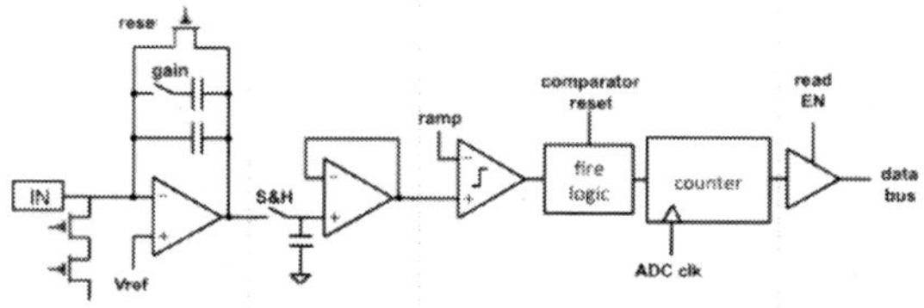

Figure 1. Schematic of a CSPAD pixel.

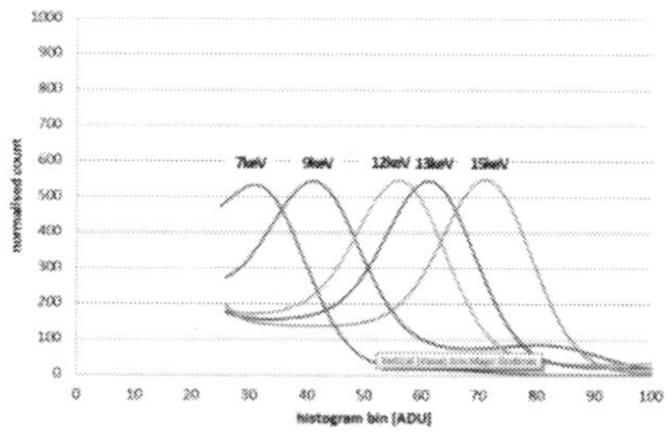

Figure 2. CSPAD 140k v1.5: energy scans.

2.2. Diagnostics at FELs

Working conditions at LCLS are very different: (i) intensity fluctuations downstream a monochromator are significant and present very peculiar intensity distributions as depicted in Figure 4 (in strong contrast to the pink beam intensity distribution); (ii) precise intensity measurements are more difficult to implement. The first step to prepare for detector tests was to improve the electronics of the systems used for diagnostics at the experimental station. Figure 5 shows a correlation plot of the beam intensity measured with the CSPAD-140k v1.5 versus the beam intensity measured with a new version of the in-house developed beam position and intensity monitor system [10]. These measurements were performed at the X-ray Pump Probe (XPP) instrument with 9 keV monochromatic radiation delivered by a silicon(111) monochromator.

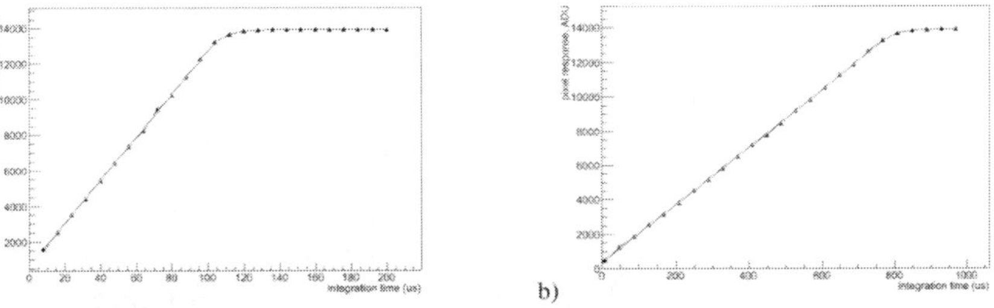

a) b)

Figure 3. Linearity measurements with the CSPAD 140k v1.2 at BL 2-2. Improved electronics resulted in good linearity both at high (a) and low gain (b).

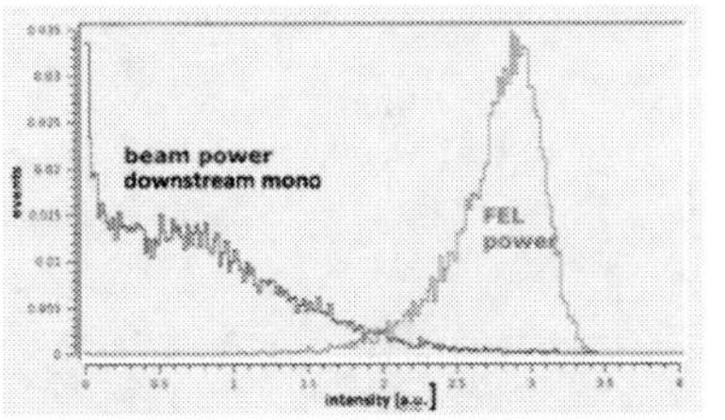

Figure 4. In red: intensity distribution measured with the gas detector. In black: intensity distribution downstream the monochromator measured with a PIPS diode (Canberra Inc.) and in-house developed readout electronics.

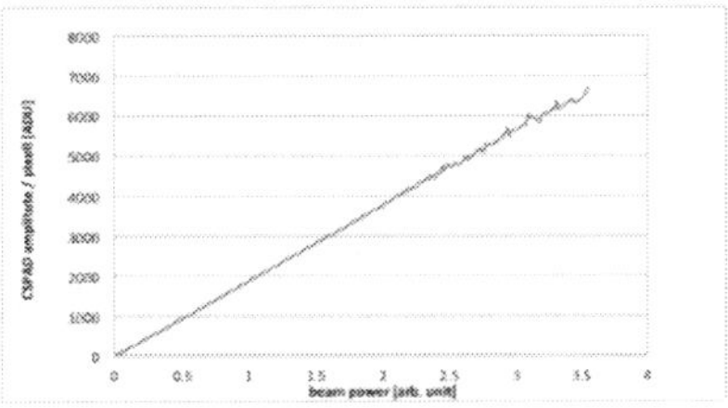

Figure 5. Correlation plots: CSPAD vs beam monitor. After optimization of the in-house electronics used to measure beam position and intensity.

2.3. Linac Coherent Light Source

To investigate how many photons arriving all at once would affect the response of the CSPAD in low gain mode, a series of measurements were taken by using small angle x-ray scattering patterns or the attenuated direct beam at 9 keV. In figure 6 are shown plots correlating the response of the CSPAD to the impinging beam power. Figure 6a shows the response of the detector with "standard" setting. Strong non-linearity can be observed at relatively low fluence. A second set of measurements aimed to extend the limit exhibited with the previous setting by changing the reference voltage of the preamplifier (figure 6b). This obviously allows extending the dynamic range but not up to the expected maximum signal of 14000 as observed previously and presented in Sec 2.1. A third study investigated the effect of the sensor bias voltage and is presented in figure 6c. We observe that for very low bias voltage the response of the detector is very similar to the one measured at SSRL (cf. figure 3).

3. Discussion

The observed behavior can be explained by taking into account the actual response of the pMOS reset and injection transistors. At high intensities and short charge collection times, the voltage builds up at the preamplifier input, due to the limited bandwidth and slew rate of the preamplifier, thus increasing the effective gate-source voltage of the reset pMOS transistor. This effectively elongates the reset time and results in some charges leaking through the still partially closed reset. To avoid this effect the signal arrival needs to be synchronized with an additional delay from the removal of the reset with respect to the value required for low intensity operation. At very short collection times the charge at the input can also forward-bias the bulk diodes of reset and injection transistors (going through the substrate in the wells and to the supply). These dynamic signal losses can be observed also in the simulations of the linearity response of the preamplifier (figure 7). The voltage on the gate of the reset pMOS transistor is assumed to switch with a 4 μs time constant equivalent to a reset release time in low signal conditions of about 3 μs. Three different collection times, 10, 100 and 200 ns, are considered (figure 7). In each case the charge arrival is supposed to happen with three possible delays: 3, 6 and 20 μs. The simulations confirm that this effect, as demonstrated by the measurements at very low biasing voltage (figure 6c), is strongly dependent on the charge collection time.

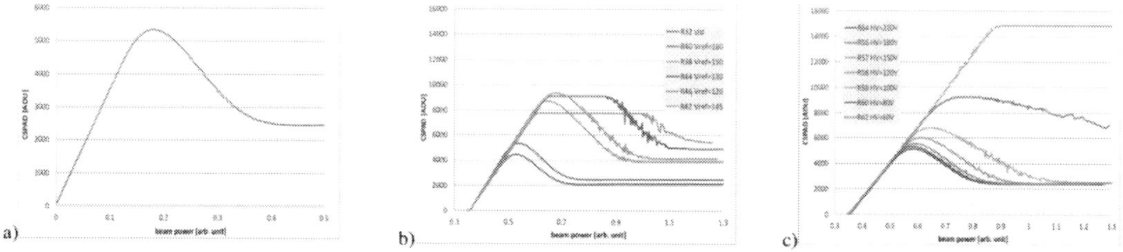

Figure 6. Linearity response of the CSPAD in low gain. a) Standard setting. b) Different reference voltages at the preamplifier. The flat-top of the response is generated by the ADCs. c) Different sensor bias voltages.

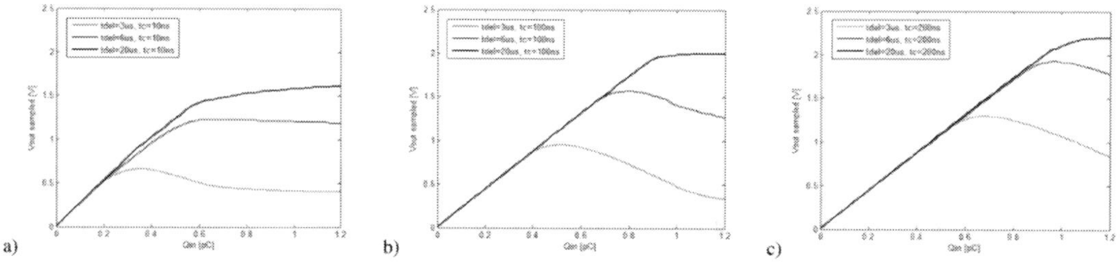

Figure 7. Simulation of the response of the CSPAD to increasing charge at the input. a) 10 ns charge collection time; b) 100 ns charge collection time; 200 ns charge collection time.

References

[1] Weidenspointner G *et al.* 2011 *Proc. SPIE* **8078** 80780U

[2] Koch A *et al.* 2013 *JINST* **8** C11001

[3] Kameshima *et al.* submitted to *Rev Sci Instrum*

[4] Koerner L J, Philipp H T, Hromalik M S, Tate M W, and Gruner S M 2009 *JINST* **4** P03001

[5] Philipp H T *et al.* 2010 *IEEE Trans Nucl Sci* **57** 3795

[6] Philipp H *et al.* 2011 *Nucl Instr Meth Phys Res A* **649** 67

[7] Hart P A *et al.* 2012 *Proc SPIE* **8504** 85040C

[8] Herrmann S C *et al.* 2013 *Nucl Instr Meth Phys Res A* **718** 550

[9] Herrmann S C *et al.* submitted to *JPCS (SRI 2013)*

[10] Herrmann S C *et al.* submitted to *JPCS (SRI 2013)*

Acknowledgments

SLAC-PUB-15641. "Portions of this research were carried out at the Stanford Synchrotron Radiation Lightsource and at the Linac Coherent Light Source (LCLS) at the SLAC National Accelerator Laboratory. SSRL and LCLS are Office of Science User Facilities operated for the U.S. Department of Energy Office of Science by Stanford University."

ePix: a class of architectures for second generation LCLS cameras

A Dragone, P Caragiulo, B Markovic, R Herbst, B Reese, S C Herrmann, P A Hart, J Segal, G A Carini, C J Kenney and G Haller

SLAC National Accelerator Laboratory, 2575 Sand Hill Rd., Menlo Park, CA, USA

E-mail: dragone@slac.stanford.edu

Abstract. ePix is a novel class of ASIC architectures, based on a common platform, optimized to build modular scalable detectors for LCLS. The platform architecture is composed of a random access analog matrix of pixel with global shutter, fast parallel column readout, and dedicated sigma-delta analog-to-digital converters per column. It also implements a dedicated control interface and all the required support electronics to perform configuration, calibration and readout of the matrix. Based on this platform a class of front-end ASICs and several camera modules, meeting different requirements, can be developed by designing specific pixel architectures. This approach reduces development time and expands the possibility of integration of detector modules with different size, shape or functionality in the same camera. The ePix platform is currently under development together with the first two integrating pixel architectures: ePix100 dedicated to ultra low noise applications and ePix10k for high dynamic range applications.

1. Introduction

The unique characteristics of Free-Electron Laser (FEL) sources in terms of brilliance and narrow pulse duration have increased the need of new large area detectors with fast readout and specifications that, depending on the experiment, can range from ultra low noise requirements [1] to extremely large full-scale and dynamic ranges [2]. The first years of the Linac Coherent Light Source (LCLS) operation have shown the need to support a wide variety of experiments with a correspondent variety of detectors with different shapes, areas and functionalities. The harsh nature of FEL sources also increases the risks of damages during operation and parts of the cameras need to be replaced often which has an impact on cost. Modular scalable detectors with standardized interfaces and communication protocols are key factors to provide time and cost effective cameras, easy to adapt and support. Only integrating detectors can be used in FEL applications and those based on architectures with a higher level of parallelism has proven to be more suitable. Hybrid pixel detectors seem at the moment the best compromise between performance and complexity [3]. Several configurations are possible within a hybrid approach and an optimal level of parallelism needs to be determined. ePix is a novel class of ASICs integrating architectures based on a common platform optimized for processing the signals from LCLS cameras. This platform is composed of a random access analog matrix of pixels with global shutter, fast parallel column readout, and dedicated sigma-delta analog-to-digital converters per column. It also implements a dedicated control interface and all the required support electronics to perform configuration, calibration and readout of the matrix. The ePix development has

been scheduled in phases. Each phase will produce a device with sufficient functionality to be used in scientific experiments. Subsequent phases will incrementally increase area and speed of the device.
Based on this platform a class of front-end ASICs and several camera modules, meeting different requirements, can be developed using different pixel architecture variants. This approach reduces development time and expands the possibility of integration of detector modules with different size, shape or functionality assembled in the same camera.
The ePix platform is currently under development together with two integrating pixel variants (table 1): ePix100, optimized for ultra low noise applications with resolution of less than 100e^- r.m.s. and a signal range of 35fC (100 photons at 8keV), and ePix10k, optimized for high dynamic range applications with resolution of less than 350e^- r.m.s. and a signal range of 3.5pC (10k photons at 8keV). The first prototype versions of the ASICs are fabricated in TSMC CMOS 0.25μm technology.

2. ePix platform architecture

The ePix platform architecture (figure 1) is composed of an analog matrix of pixel with global shutter, fast parallel column readout, and dedicated sigma-delta analog-to-digital converters per column.

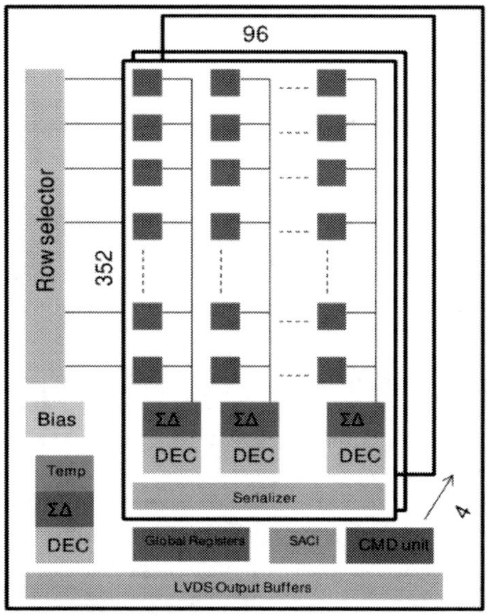

Figure 1. Simplified block diagram of the ePix Platform (number of rows and column reported are for ePix100).

Table 1. ePix ASICs characteristics

	ePix100	ePix10k
Pixel size	50x50 μm^2	100x100 μm^2
Array size	352x384	176x192
Frame rate	500Hz (min. req. 120Hz)	1kHz (min. req. 120Hz)
Signal range	220ke^-	22Me^-
ENC	50e^- (min. req. 100e^-)	120e^- (min. req. 350e^-)
Gains	6.5μV/e^-	6.5μV/e^-, 64nV/e^-
DC Power consumption	10μW/pix	10μW/pix

Figure 2. ePix 2x2 camera housing concept.

Considering the limited speed and complexity of ADC architectures that could be implemented in pixels, having an ADC per column, optimizes the performance in terms of conversion and readout time. It also allows the design of fully analog pixels and frees up space in the pixels that can be used for more advanced filtering circuitry. The choice of Sigma Delta converters [4] is a trade-off among sampling rate, resolution, level of digital multiplexing, serialization rate and development time. Sigma Delta converters have the possibility to control the resolution adjusting the oversampling rate, and thus trading resolution for conversion speed. Their intrinsic pseudo-random nature reduces amplitude dependent cross-talk among channels in systems where several ADCs operating at the same time are present. The architecture also implements a dedicated serial control interface that allows handshake as opposed to standard SPI or I2C interfaces. All the required support electronics to perform configuration, calibration and readout of the matrix are also implemented. Columns are divided in 4 banks. To achieve maximum frame rate the output of all ADCs in a bank will be serialized, encoded and transmitted on a single high speed LVDS link. The 4 banks are readout in parallel to speed up the readout process. The acquisition and the readout phase are non-overlapped and are controlled by a single periodic signal (ACQ) whose period defines the integration time. This signal is synchronized to the LCLS beam trigger. During the active part of the ACQ period the charge from the pixels is integrated and sampled. During the inactive period stored data are readout and the system is reset. A power pulsing method is implemented, reducing the power consumption of the various sections of the ASIC during their respective idle states. Support electronics to perform gain calibration per pixel is included together with the possibility to track and correct common mode variations between frames and temperature variations.

3. ePix100 and ePix10k pixel architectures

Figures 3 and 4 respectively show simplified block diagrams of the pixel architecture of ePix100 and ePix10k.

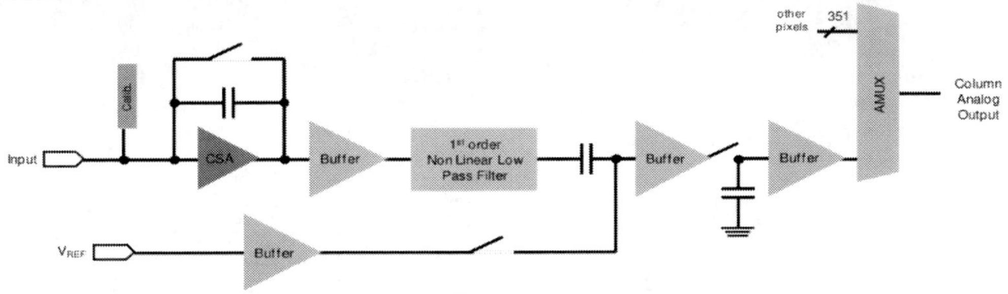

Figure 3. Simplified block diagram of the ePix100 pixel.

Each pixel implements a single stage low noise charge integrator with a pulsed reset, a first order non-linear programmable Low Pass (LP) filter, a correlated double sampler (CDS) and a sample and hold stage followed by a column buffer. Because of the limited area the circuits in the blocks are not complex and small buffers are required to decouple the various sections.

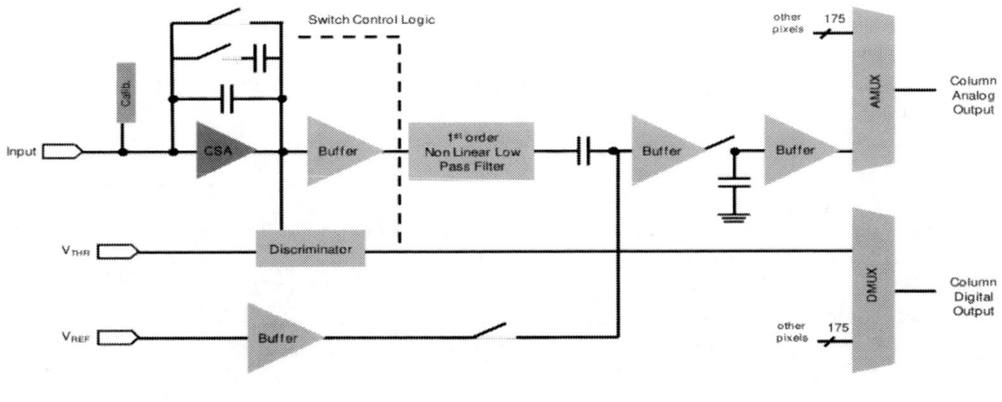

Figure 4. Simplified block diagram of the ePix10k pixel.

In ePix100 noise requirement is dominant and thus noise considerations drive the optimization of the signal chain. In ePix10k the large signal range is the driving requirement. ePix10k implements multiple gain settings and auto-ranging [5]. The auto-ranging system allows reducing dynamically the gain of the pre-amplifier depending on the input signal level and thus extending the dynamic range without sacrificing the resolution for small signals. The first order LP filter is designed using active resistors to limit the area taken. Active resistors are nonlinear but are optimally sized to produce maximum filtering at low number of photons. The time constant of the filter is continuously programmable by an internal DAC. The CDS stage is used to sample and store the filter output, eliminating the fluctuations due to KTC noise and baseline fluctuations, and to reduce low frequency noise [6]. In this architecture a "quasi-trapezoidal" noise weighting function can be implemented. Figure 5 and 6 show the layout of the two different pixel designs.

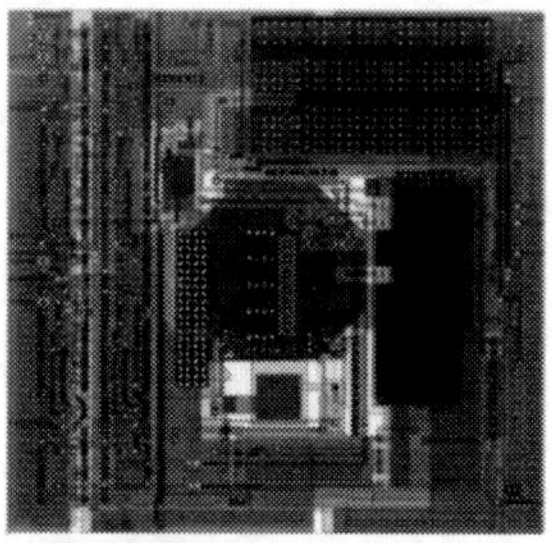

Figure 5. ePix100 pixel 50μmx50μm

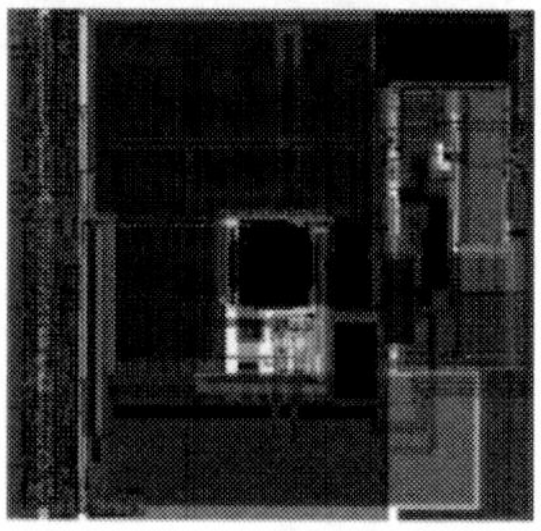

Figure 6. ePix10k pixel 100μmx100μm

Figure 7. ePix100 96x96 pixel prototype.

The first prototype of ePix100 (figure 7) has been fabricated and tests to assess the performance are underway.

Acknowledgments
SLAC-PUB-15640. Use of the SLAC National Accelerator Laboratory, was supported by the U.S. Department of Energy, Office of Science, Office of Basic Energy Sciences, under Contract No. DE-AC02-76SF00515.

References

[1] Hruszkewycz S O, Sutton M, Fuoss P H, Adams B, Rosenkranz S, Ludwig K F Jr, Roseker W, Fritz D, Cammarata M, Zhu D, Lee S, Lemke H, Gutt C, Robert A, Grübel G, Stephenson G B 2012 *Physical Review Letters* **109** 185502
[2] Boutet S, Lomb L, Williams G J, Barends T R M et al. 2012 *Science.* **337** 362
[3] Hart P, Boutet S, Carini G A, Dragone A, Duda B, Freytag D, Haller G, Herbst R, Herrmann S, Kenney C, Morse J, Nordby M, Pines J, van Bakel N, Weaver M, Williams G, 2012 *NSS-MIC Conference records.* 538
[4] Schreier R, Themes G C 2004 *Understanding Delta-Sigma Data Converters* (New York: Wiley-IEEE Press)
[5] Freytag D, Herbst R, Brau J, Breidenbach M, Frey R, Haller G, Holbrook B, Lander R, Nelson T, Radeka V, Strom D, Tripathi M, 2008 *NSS-MIC Conference records.* 3447
[6] Porro M, Fiorini C, Strüder L 2003 *Nucl. Instrum. Methods, A.* **512** 179

Journal of Physics: Conference Series **493** (2014) 012013 doi:10.1088/1742-6596/493/1/012013

CSPAD upgrades and CSPAD V1.5 at LCLS

S Herrmann, P Hart, A Dragone, D Freytag, R Herbst, J Pines, M Weaver, G A Carini, J B Thayer, O Shawn, C J Kenney, G Haller

SLAC National Accelerator Laboratory, 2575 Sand Hill Rd, Menlo Park, 94025 CA

E-mail: herrmann@slac.stanford.edu

Abstract. After improving the PCB level electronics the next step in our continuing upgrade program of LCLS Cornell-SLAC Pixel Array Detector (CSPAD) cameras is the use of a new improved ASIC named CSPAD V1.5. The upgraded ASIC includes on chip DACs to set the bias currents of all amplifiers. The new chip also supports power cycling by design. Together this simplifies the PCB level complexity and eases the integration of many ASICs into one camera. Homogeneity across the full reticle size chip was improved by redesigning the power distribution. The upgrade included modifications of the gain latches and the per pixel comparators. Results from the upgraded cameras used at LCLS will be presented and discussed.

1. Introduction

The Cornell-SLAC Pixel Array Detector (CSPAD) [1 - 5] is the workhorse for LCLS hard x-rays experiments. After deploying three 2.3Mpixel and many 140kpixel cameras, SLAC detector group has focused in improving the detector performance and optimizing its use in experiments. Upgrading electronics, firmware and ASIC was needed to improve the performances of these detectors. In addition refining characterization, calibration and operating conditions allowed understanding fine details and finally optimizing the cameras for a given experiment. The upgrades produced first a new version defined as CSPAD V1.2, which implements new electronics, firmware and mechanics, and then, with the addition of a new ASIC, the CSPAD V1.5, which will be the last of this family. Second generation cameras will be built on a new SLAC platform, the ePix family [6]. The evolution of the cameras and their use it is shown in Fig. 1. The basic characteristics of the CSPAD are summarized in Table 1. Upgrades and experimental results are discussed in the following paragraphs.

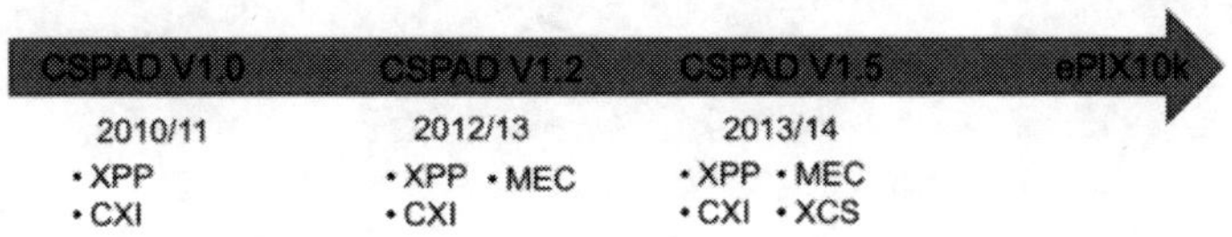

Figure 1. CSPAD cameras: upgrades and use at LCLS.

2. Design upgrades

The first set of upgrades targeted electronics, firmware and some mechanical components. We designed a set of new PCBs with modified power supply and improved bias and ramp generation circuits. These PCBs were swapped in during instrument maintenance at CXI and XPP. The new

Journal of Physics: Conference Series **493** (2014) 012013 doi:10.1088/1742-6596/493/1/012013

electronics have helped mitigate the nonlinearity and crosstalk problems observed with the first version of the cameras [5]. The cameras with the improved electronics are labeled as version 1.2. Another important upgrade involved power supplies and cable harness. Besides a significantly improved operational reliability, the new power supply units can support one 2.3Mpixel together with up to four 140kpixel cameras for experiment with concurrent techniques using multiple detectors. Alternatively up to eight CSPAD-140k detectors can be operated (this variant is deployed at the MEC instrument).

Mechanics upgrades were implemented to improve the thermal stability and enable a more compact camera housing for the 2.3Mpixel XPP detector. A new compact solid cooling plate replaced the original CXI-style design. This is especially important because this camera is mounted in a robotic arm and the very compact housing is a premium for placing the detector in critical positions around the sample.

Firmware and Software are continuously improved. The camera front-end firmware has seen improvements for debug-ability and slight performance increases. It now also includes the option to integrate for a full 8ms cycle, enabling the use of the camera at synchrotron with less than 1% dead time. The data acquisition software used for configuring and reading out the CSPAD has been made more fault-tolerant and efficient. Support has also been added for compressing CSPAD data. Many features have been added to improve analysis of the CSPAD image in real time, including: a hit finder that can also do thresholding and pixel sums, image averaging for binned and unbinned images, certain geometry corrections, improvements in common mode calculation, center of mass calculation, calculation of mean and variance of image, and the ability to suppress or enable pedestals at the click of a button. The software is highly configurable for specific experimental conditions and more responsive to user input.

Table 1. CSPAD properties

Pixel size	110 μm x 110 μm
Area	326 cm^2 (2.3 Mpixel)
Maximum signal	2700 8keV photons/pixel (low gain) 350 8keV photons/pixel (high gain)
Frame rate	120Hz
Noise	~ 3.5 keV (low gain), ~ 1 keV (high gain)

Finally the CSPAD ASIC was modified to overcome some of the limitations experienced with the first version of the chip. Without significant changes in the architecture, the upgraded ASIC includes on chip DACs to set the bias currents of all amplifiers. The new chip also supports power cycling by design. This simplifies the PCB level complexity and eases the integration of many ASICs into one camera. Homogeneity across the full reticle size chip was improved by redesigning the power distribution. The improved ASIC homogeneity across the full chip is shown in Figure 2.

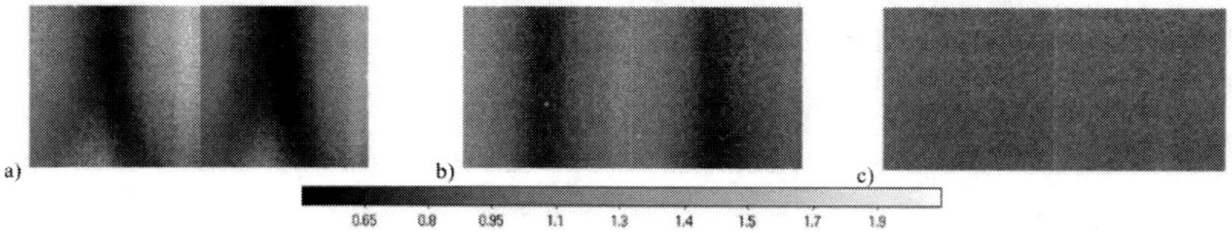

Figure 2. Series showing the improvements in homogeneity of different CSPAD versions. a) V1.0, b) V1.2 and c) V1.5. The images show the noise distribution of a CSPAD building block consisting of two ASICs bump bonded to one sensor element. The images (in arbitrary units) are normalized to show the same average noise level (set to 1.0). The scale includes the range from 50% to 200% of the average noise level of the module.

3. Measurements and camera deployment

All these upgrades made the CSPAD detectors easier to use and to adjust to the needs of very different experiments with requirements within the limit of these cameras. Performance improvement has been systematically investigated and validated with x-rays at synchrotrons and at LCLS [7]. For example figure 3 shows the improvement on the linearity of the 2.3Mpixel camera. The measurements were carried out at the Stanford Synchrotron Radiation Lightsource (SSRL) BL 2-2 with 9keV monochromatic beam by increasing the integration time under constant illumination.
The improved homogeneity of the CSPAD V1.5 ASIC can be also be observed in the spectroscopic measurements shown in figure 4. The measurements were taken at SSRL BL 7-2. The histogram shows the CSPAD response to five monochromatic energies. All pixels of the four ASICs comprised in the 140kpixel camera are included in this histogram without any pixel gain-correction applied. All the measurements were performed at room temperature.

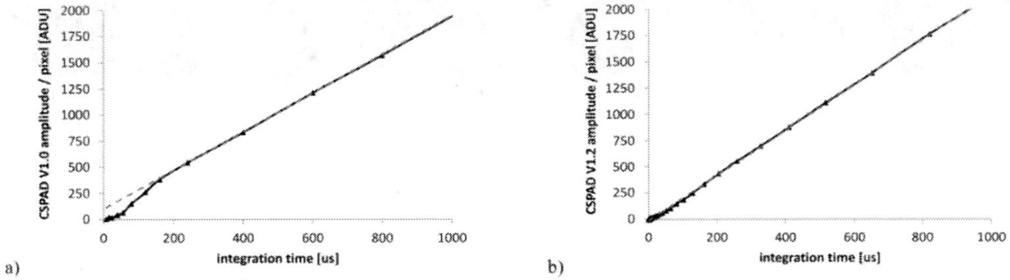

Figure 3. Linearity response of CSPAD in high gain mode. Measurements carried out at SSRL BL 2-2. To achieve a linear input signal the integration time of the camera was changed under constant illumination conditions. The early version of the 2.3Mpixel CSPAD detector (a) shows a distinct nonlinearity at low signal levels caused by pixels crosstalk. The improved camera version 1.2 delivers better linearity response (b).

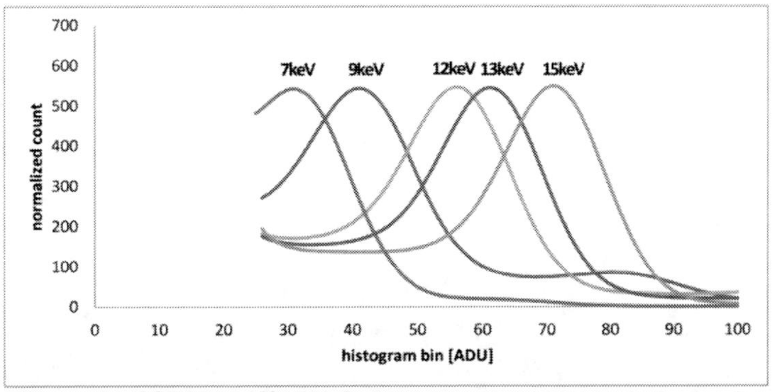

Figure 4. CSPAD-140k V1.5: histograms of all the pixels for five different monochromatic energies. Measurements performed at SSRL BL 7-2. The histograms only include entries with measured amplitudes higher than 25 ADU. Counts have been scaled on the y-axis for normalisation to show the same peak height for comparison. No pixel gain correction has been applied.

As result of this continuous effort to improve CSPAD detectors and user interface, these cameras have been used in all the hard x-rays instruments at LCLS in a variety of experiments. Thanks to a better control of the detector settings, challenging measurements at the limit of the detector capability, e.g.

low noise, low signal, low energy (as low as 2.2keV) and very high maximum signal over large areas, were successfully performed. In addition CSPAD-140ks have been used at SSRL, SACLA and at the Advanced Photon Source (APS). Examples of setup are shown in figures 5 and 6.

Figure 5. Arrangement of multiple CSPAD-140k cameras (left) including one used in a Thomson spectrometer (right) at the MEC instrument of LCLS.

Figure 6. The 2.3Mpixel CSPAD V1.2 used for an experiment at XCS: the large area and fast readout were crucial for a delicate alignment.

References

[1] Koerner L J, Philipp H T, Hromalik M S, Tate M W, and Gruner S M 2009 JINST 4 P03001
[2] Philipp H T et al. 2010 IEEE Trans Nucl Sci 57 3795
[3] Philipp H et al. 2011 Nucl Instr Meth Phys Res A 649 67
[4] Hart P A et al. 2012 Proc SPIE 8504 85040C
[5] Herrmann S C et al. 2013 Nucl Instr Meth Phys Res A 718 550
[6] Dragone A et al. submitted to JPCS (SRI 2013)
[7] Carini G A et al. submitted to JPCS (SRI 2013)

Acknowledgments

SLAC-PUB-15694. "Portions of this research were carried out at the Stanford Synchrotron Radiation Lightsource and at the Linac Coherent Light Source (LCLS) at the SLAC National Accelerator Laboratory. SSRL and LCLS are Office of Science User Facilities operated for the U.S. Department of Energy Office of Science by Stanford University."

Diode readout electronics for beam intensity and position monitors for FELs

S Herrmann, P Hart, M Freytag, J Pines, M Weaver, L Sapozhnikov, S Nelson, J Koglin, G A Carini, A Tomada, G Haller

SLAC National Accelerator Laboratory, 2575 Sand Hill Rd, Menlo Park, 94025 CA

E-mail: herrmann@slac.stanford.edu

Abstract. LCLS uses Intensity-Position Monitors (IPM) to measure intensity and position of the FEL x-ray pulses. The primary beam passes through a silicon nitride film and four diodes, arranged in quadrants, detect the backscattered x-ray photons. The position is derived from the relative intensity of the four diodes, while the sum provides beam intensity information. In contrast to traditional synchrotron beam monitors, where diodes measure a DC current signal, the LCLS beam monitors have to cope with the pulsed nature of the FEL, which requires a large single shot dynamic range. A key component of these beam monitors is the readout electronics. The first generation of beam monitors showed some limitations. A new scheme with upgraded electronics, firmware and software was implemented resulting in a more robust and reliable measuring tool.

1. Introduction

LCLS [1] uses Intensity-Position Monitors (IPM) to measure intensity and position of the FEL x-ray pulses [2][3]. The primary beam passes through a silicon nitride film and four photodiodes, arranged in quadrants, detect the backscattered x-ray photons. The position is derived from the relative intensity of the four diodes, while the sum provides beam intensity information. In contrast to traditional synchrotron beam monitors where diodes measure a DC current signal, the LCLS beam monitors have to cope with the pulsed nature of the FEL. This requires dedicated pulse processing electronics, called Intensity Position Intensity Monitor Board (IPIMB), which has to quantify a large dynamic range of charge signals on a shot-by-shot basis. The original design and the modifications implemented in the upgraded version are presented and discussed in this manuscript.

2. Design

The first generation IPIMB electronics for reading out these diodes was deployed in 2010 and used a charge sensitive amplifier followed by a RCCR shaper. The charge sensitive amplifier is AC coupled and the photodiode bias voltage is applied over a large biasing resistor. Due to the known arrival time of the FEL pulse, the time when the waveform produced by the RCCR shaper reaches its maximum is also known. Therefore the amplitude information can be achieved by sampling the signal amplitude with a single ADC sample at this time, without an additional peak stretcher. Baseline fluctuations at the output of the filter, generated for instance by fluctuations of the FEL beam power, are corrected by taking eight samples before the FEL pulse and subtracting their average from the sampled peak. After each FEL pulse the resistive bias network has to recharge the diode. This results in a slowly rising

voltage over the diode whose steepness depends on the average beam energy within the time constant of the bias network. The rising voltage in turn offsets the output of the simple RCCR shaper causing additional systematic variations of the baseline. In addition, changes in environmental conditions, such as temperature and stray light, affect the diode properties.
Various mechanisms limit the performance of this analog processing scheme. As the amount of charge detected should ideally cover up to five orders of magnitude to make the device a usable diagnostic under real experimental conditions, the rise time and shape of the signal at the diode and after the preamplifier varies considerably. The RCCR shaper imposes a compromise between the practical shaping time and the sensitivity to rise time variations. Furthermore, the signals are affected by pickup over the relative long leads needed to connect the diodes, that are mounted inside a vacuum chamber on the beamline, to the electronics in air. This pickup is often a combination of various but specific frequencies. The RCCR filter attenuates high frequency noise but does not suppress selected frequencies. As only one output value is measured, it can be very difficult to identify the origin of a possible pickup from beamline components such as a vacuum pump or motorized stage.

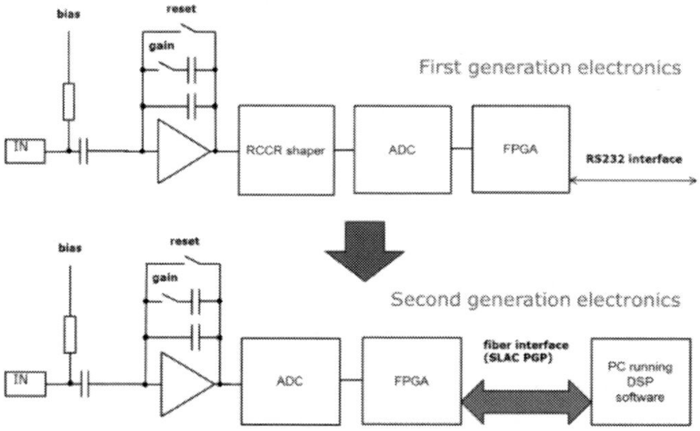

Figure 1. Block diagrams of the first (top) and second (bottom) generation electronics for photodiode readout at LCLS. The second generation replaces the fixed RCCR shaper with software-based digital signal processing of the sampled waveform.

The original design was modified by removing the shaper and instead sampling the waveform at the output of the charge sensitive amplifier continuously. All waveform samples are sent to the data acquisition system via a fiber optical high speed link (SLAC Pretty-Good-Protocol, PGP). The sampled waveform can then be shaped in real-time applying digital signal processing methods available within the online monitoring software. The software is capable of waveform fitting or Finite Impulse Response (FIR) filtering. The block diagrams of the two systems, original and modified, are shown in figure 1.
Figure 2 (top) shows a representative waveform response of an x-ray photodiode to a photon pulse. It is worth mentioning that the rise time of the signal is slightly dependent on the amplitude and gain settings. This can be attributed mainly to the limited slew rate and current capability of the op-amp used in the charge sensitive amplifier. In fact, the large instantaneous signals on the diode impose challenging requirements for the front-end amplifier. To be less sensitive to these variations and suppress baseline fluctuations, we used a simple FIR trapezoidal filter with flat top. This filter has also the advantage of exhibiting distinct zeros in the frequency transfer function: specific pickup frequencies can be supressed by choosing the right averaging length. Given a dataset of example waveforms, an optimised filter can be synthesised to maximise the signal-to-noise ratio. Different filters optimised for different operation regimes (low, medium, high signal) could also be considered. The FIR filter coefficients can be loaded into the online monitoring software and the waveform

processed in real-time for all the LCLS pulses (120 events per second). The lower half of figure 2 shows an example of the resulting waveform after filtering. The software can then use the peak of the filtered waveform as a scalar value of the diode signal for further processing.

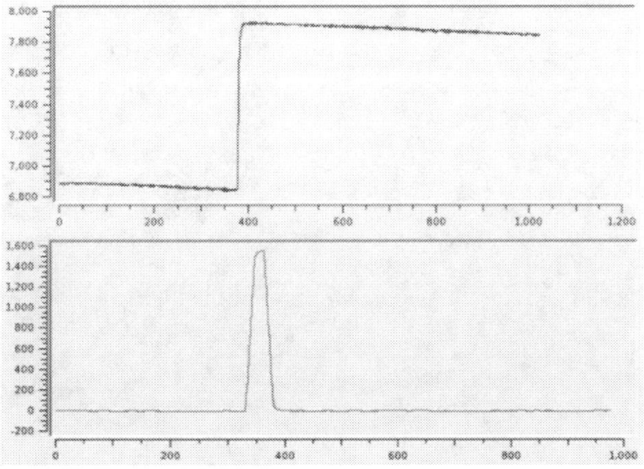

Figure 2. Example of a captured waveform from a photodiode (top) as displayed in the online monitoring software. The software applies an arbitrary FIR filter to the waveform in real-time (bottom). The peak of this filtered waveform can be used as a scalar value representing the diode signal for further processing.

Having access to the amplifier output waveform in real-time from the online monitor helps considerably with debugging an experimental setup: parameters such as timing, signal amplitude and the recovery time from the diode bias network can be immediately seen. Furthermore the waveform can be analysed for specific pickup frequencies and the experimental setup investigated for possible noise sources accordingly. This new system, Intensity Monitor with SLAC PGP (IMP), has 4 channels and is designed to readout signals from photodiodes or avalanche photodiodes. The calculated dynamic range of this electronics is up to 3 million photons with a noise of 25 photons at 8 keV photon energy when used with an x-ray photodiode. As shown in the following section, the IMP can also be combined with a commercial Silicon Drift Detector (SDD).

3. Measurements and Conclusions

Measurements were performed at the CXI instrument [4] at LCLS using pink SASE beam at 9 keV. The primary beam hit a sample (100 µm glass with 50 nm gold coating) producing roughly uniform scattering and a small fraction of the photons from the primary beam were impinging on the diode. Figure 3 shows the IMP box and the commercial x-ray photodiode used for this test.

In figure 4, as a proof of concept, the signal measured with the diode is plotted against the FEL beam energy measured with the LCLS gas detector [5]. The x-axis spans the entire range of the FEL pulse energy used during this measurement.

To demonstrate the flexibility of the approach we connected the IMP electronics to a commercial SDD which includes a preamplifier with pulsed reset. The captured waveforms were processed with a trapezoidal FIR filter very similar to the one used for the photodiode. The resulting spectrum, a histogram of the measured signal amplitudes over many FEL shots, is shown in figure 5. The achieved energy resolution is 180 eV for the Iron K_α line and 230 eV for the simultaneous pileup of two Iron K_α photons. These results show that this new system (IMP) addresses the issues of the previous version (IPIMB) and is suitable for a variety of applications.

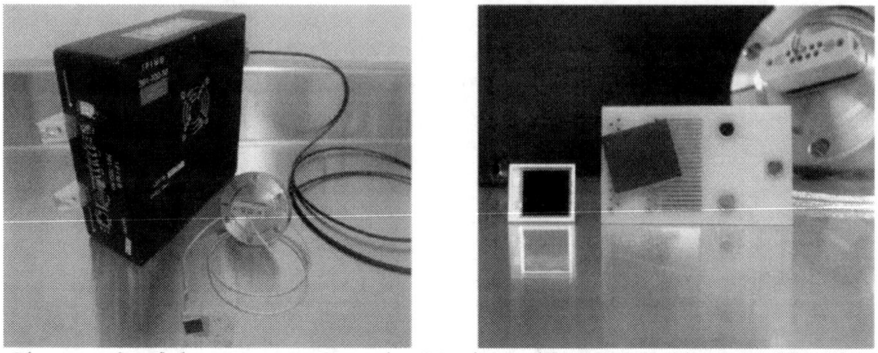

Figure 3. Photograph of the prototype Intensity Monitor with PGP (IMP) to readout photodiodes (left). The photograph on the right side show the commercial x-ray photodiode used for the measurements. A sheet of 30 μm Kapton is glued in front of the diode to block optical light.

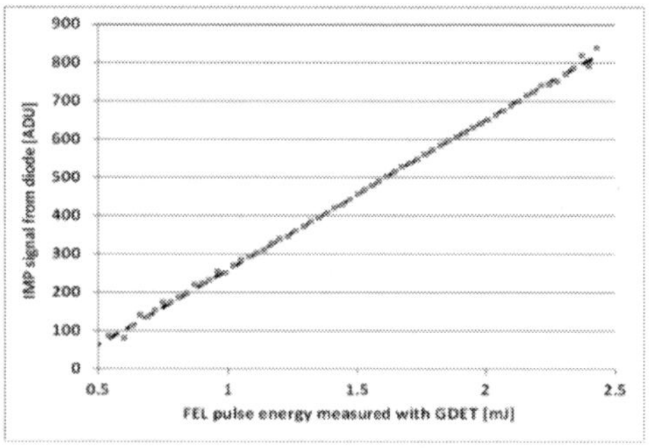

Figure 4. Diode signal vs. LCLS beam pulse energy measured with a gas detector. The measurement was performed at the CXI instrument of LCLS with pink SASE beam at 9 keV. The x-axis spans the full pulse energy range of the FEL used during this measurement.

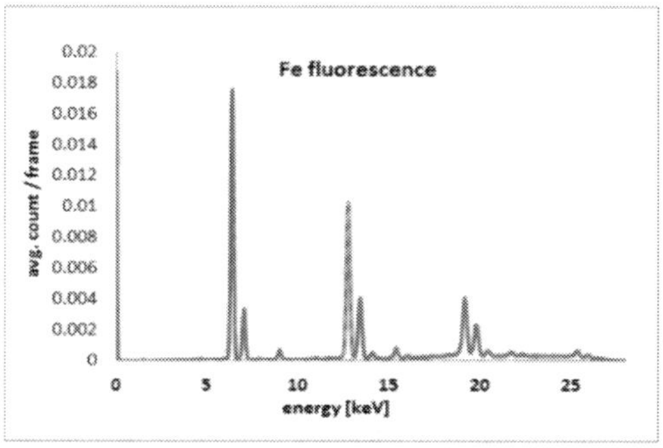

Figure 5. Iron fluorescence spectrum measured with the IMP prototype connected to an SDD. For single photon events the Iron K_α, K_β and the primary beam energy of 9 keV can be identified. Further two, three and four photon pileup events can be seen in the spectrum. The measurement reached 180 eV energy resolution on the single photon Iron K_α line.

References

[1] Emma P *et al.* 2010 *Nature Photonics* 4, 641 - 647
[2] Feng Y *et al.* 2011 *Proc SPIE* 8140 81400Q
[3] Tono K, Kudo T, Yabashi M, Tachibana T, Feng Y, Fritz D, Hastings J, Ishikawa T 2011 *Rev. Sci. Instrum.* 82 023108
[4] Boutet S, Williams G 2010 *New J. Phys.* 12 035024
[5] Hau–Riege S P, Bionta R M, Ryutov D D, and Krzywinski J 2008 *J. Appl. Phys.* 103, 053306

Acknowledgments

The authors wish to thank the CXI and XPP staff for their supports during the measurements.
SLAC-PUB-15695. "Portions of this research were carried out at the Stanford Synchrotron Radiation Lightsource and at the Linac Coherent Light Source (LCLS) at the SLAC National Accelerator Laboratory. SSRL and LCLS are Office of Science User Facilities operated for the U.S. Department of Energy Office of Science by Stanford University."

Compact pseudo-2D strip detector system for sub-meV IXS

J W Keister[1], D P Siddons[1], A Kuczewski[1], D S Coburn[1], R Beuttenmuller[2], A Rumaiz[1] and Y Q Cai[1]
[1]Photon Sciences Directorate, Brookhaven National Laboratory, Upton, NY 11973-5000 USA
[2]Instrumentation Division, Brookhaven National Laboratory, Upton, NY 11973-5000 USA

E-mail: jkeister@bnl.gov

Abstract. A dual 32 strip sensor array is realized for use at the IXS (inelastic x-ray scattering) beamline of NSLS-II. By making use of established controls methods and sensor device recipes, our new geometry is realized quickly and at minimal cost. The detector geometry is chosen to match the output of a multi-element high-resolution energy analyzer, while the pulse thresholding is optimized for an ultra-low noise floor at a pass energy of 9.13 keV. Detector subsystems and integration are described, including sensor geometry and silicon device processing, cooling and thermal readback, bias and threshold optimizations, readout ASIC and controls, vacuum enclosure and x-ray window, mounting and positioning, and assembly procedure and testing.

1. Introduction

Inelastic x-ray scattering (IXS) presents unique challenges for detector development. In backscattering geometry, the detector is typically located near the sample, where space is limited; this constraint is compounded by the need for multiple sensor elements to support multiple analyzers or segments. Furthermore, in many IXS experiments where countrates can be particularly low, electronic noise and cosmic background must be fully suppressed.

Utilizing in-house capabilities for mechanical and electronics fabrication, a custom multi-pixel sensor system was undertaken to support the IXS beamline of NSLS-II [1] which makes use of a novel grazing-incidence backscatter scheme to achieve sub-meV resolution at 9.13 keV photon energy [2]. With first light anticipated in 2014, a rapid and low-cost detector development was pursued, providing the unique capabilities needed utilizing established detector capabilities at BNL [3].

2. Requirements

The geometrical requirements for the IXS detector system derive from the spectrometer optics, as illustrated in figure 1. In the IXS beamline, 5-10 mrad of scattered x-rays are collected from the sample using a multilayer mirror at 200 mm distance, which generates a 1-2 mm beam, collimated to within 100 μrad [4]. This collected beam is incident upon a set of three types of silicon crystals which includes collimator (C), dispersion element (D) and wavelength selector (W). The C crystal, reflecting asymmetrically, delivers an output beam of 21.5 mm in vertical extent per mm of incident beam size. Since a D crystal of 200 mm length, reflecting via grazing incidence back-diffraction, can accept no more than 1.2 mrad of equivalent collection from the mirror, the IXS beamline analyzer is designed to carry up to six D crystals in line, for a total acceptance of 7.3 mrad. After the D crystal reflection,

Journal of Physics: Conference Series **493** (2014) 012015 doi:10.1088/1742-6596/493/1/012015

beam passes through the C crystal via anomalous transmission and is restored to its original size by the W crystal before finally striking the detector. This arrangement is designed to produce 0.7 meV bandwidth [1]. The detector's energy resolution therefore is less critical than its geometry and thresholding features.

Lengthening the effective D crystal surface in this way yields an arrangement for the crystal analyzer assembly which is isomorphic to a dragonfly — with segmented abdomen, W as head, and C as thorax — so the IXS analyzer also bears this nickname. Since the reflection of each D crystal is independent, it is necessary that the detector separates the "beamlets" delivered from each.

As the IXS spectrometer is designed to support up to five such "dragonfly" analyzers, arranged side by side (multiple q option), a maximum of ~200 mm of horizontal extent is available for a detector system at each analyzer output location. The detector system must also avoid interference with beam incident to the analyzer, so it is constrained in vertical extent as well. Further requirements for the IXS detector system include high efficiency (>90%), low noise rate (<0.01 cps), energy resolution ($\Delta E/E$) < 10%, and maximum count rate >20 kcps.

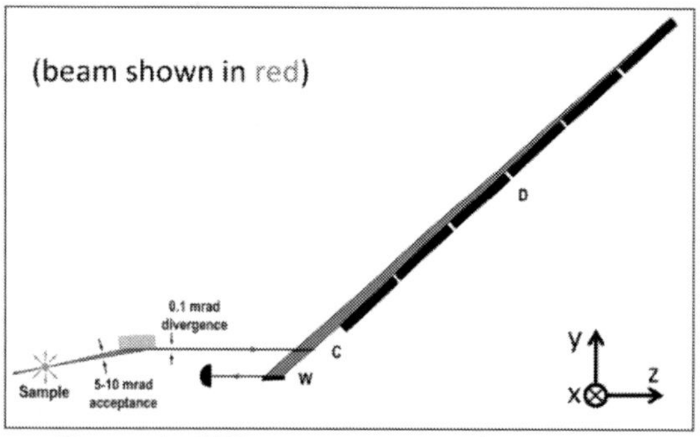

Figure 1. IXS spectrometer optical concept.

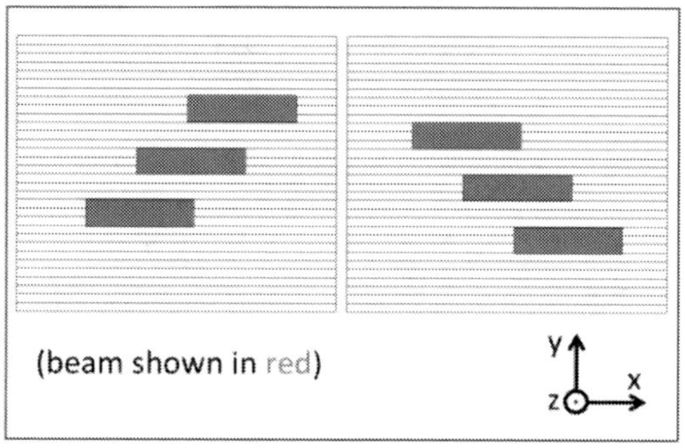

Figure 2. "Tagma" detector concept.

3. Implementation

Realization of the IXS Tagma detector system involves several subsystems as well as system assembly and test. Planning and execution of each element is briefly described in the following sections.

3.1 Sensor Geometry and Fabrication

To address the required isolation of the IXS analyzer beamlets, a custom sensor geometry was developed. A goal of no less than one unused sensor channel between adjacent beamlets was desired. Since the accepted 7.3 mrad is to be contiguous, a horizontal angle (ϕ) offset must be applied to each D crystal in order to ensure that adjacent beamlets do not share sensor elements. Consistent with this approach is the theoretical expectation that horizontal offset of 1±0.5 mrad is ideal for avoiding multiple beam losses near exact backscattering, with negligible energy shift [5]. Consequently, a total D-W-detector distance of 1 m results in a horizontal deflection of 2 mm at the detector plane. Likewise, shorter distances produce proportionally smaller offsets. By alternating the direction of the horizontal steering from one D crystal to the next, vertically adjacent beamlets can be made to land on either of two one-dimensional strips patterned side by side, as illustrated in figure 2. As each strip is 3 mm wide, a vertical pitch of 80 μm is selected to minimize interpixel capacitance (crosstalk and noise) while offering spatial resolution sufficient for the beamlet size (~244 μm V×1 mm H). Each beamlet is expected to occupy ~3 sensor channels (for 18 total) and to be separated by its neighbors by either unused channels or the central ground line, for maximum isolation. This pair of adjacent strips can be referred to as "pseudo-2D." In the spirit of the "dragonfly" analyzer we also refer to this detector as "Tagma," owing to its specialized grouping of segments into a coherently functional unit.

For maximum efficiency, the sensors were made from silicon wafers of 400 μm thickness (98% efficient at 9.13 keV), using an established P-on-N recipe. The IXS sensor geometry is included in a set of 6" wafer masks produced for a collection of similar sensors, including simple diodes and 640-

strip arrays. All process steps aside from ion implantation were performed in the cleanrooms of BNL Instrumentation. Aluminum window metallization thickness was 200-250 nm (>99.7% transmissive). Microscope inspection as well as I-V and continuity characterization was performed on candidate sensor chips to select units for system integration. The sensor chips include unpatterned areas for mechanical and thermal support.

3.2 Vacuum and Mechanical Assembly

Taking advantage of existing methods in high vacuum design, a set of three major mechanical parts were developed: a vacuum body with conflat nipple for mounting to the analyzer chamber, a cover plate with water-cooling and electrical feedthroughs, and a water-cooled copper block to support the sensor board. These three main parts were produced by BNL central fabrication facilities, with features added, complementary parts fabricated, and assembly completed by beamline technical staff. The copper block is supported from the cover plate in vacuum, and supports the sensor and electronics board; a rectangular hole is also included to let the x-rays pass into the sensor window. The vacuum body includes an o-ring groove (for 1/16" #47 size o-ring) to provide sealing to the cover. The cover plate includes compression fittings for water cooling tubes and a 37-pin electrical feedthrough (Detoronics DRTBH4-37P with 1/16" #29 size o-ring). An illustration of the assembly is provided in figure 3. Not included in this figure are the in-vacuum cables which fit tightly between the cover plate and sensor board; a conductive shield is also included to protect the sensor and wirebonds. The constraints on available space for the detector assembly for the IXS spectrometer have led to a design with minimum clearances (as small as ~2 mm) between components.

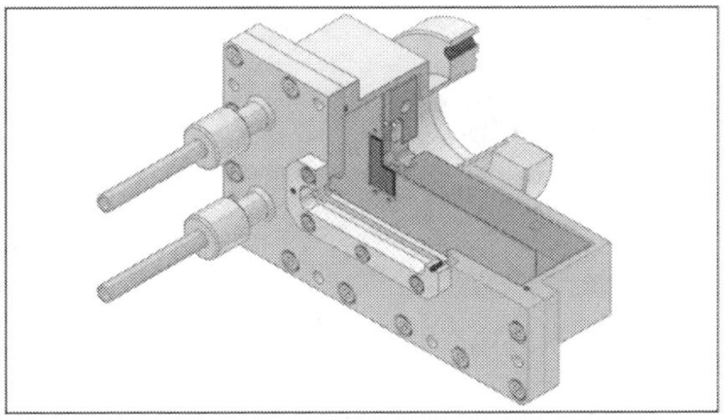

Figure 3. Tagma detector assembly (section view).

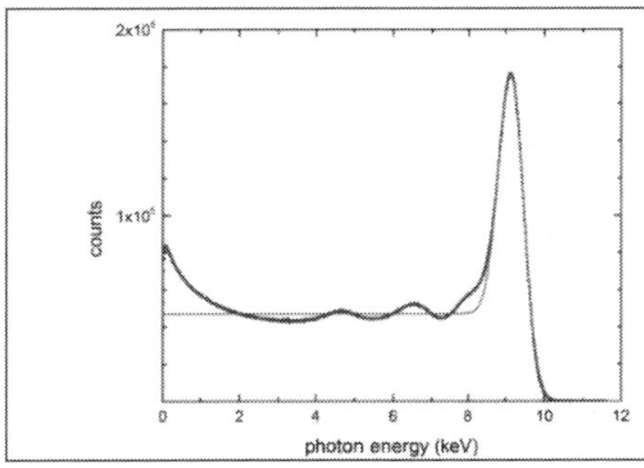

Figure 4. Spectrum for 9.1 keV x-rays.

The copper water cooling block was assembled using soft solder and tested to 1 MPa (150 PSI) without failure, for use at 700 kPa (100 PSI). For assembly of the sensor board to the copper block, an intermediate copper sensor support frame is used, which supports the sensor chip atop a pair of thermoelectric Peltier coolers (TE Technology TE-2-(31-12)-1.0, wired in series) which in turn rest on the water-cooled block. The combined height of coolers and small frame are matched by the depth of the mating surface within the large block in order to maintain the sensor and board at the same height. A pair of 10 kΩ thermistors attached to both copper parts offer temperature readback for the sensor and bath. Good thermal contact at interfaces (sensor to frame to Peltier to copper block, thermistor to frame and copper block) is ensured using a high-resistivity, high-thermal conductivity epoxy (EG-7658), cured for 2 hours at 100 °C.

3.3 Board Design and Integration

Sensor readout is provided via the HERMES 4 ASIC [3] which amplifies, filters and counts photon pulses, with 32 channels per chip. It can count up to 10^5 pulses per second per channel, and offers both upper and lower thresholds. A custom board was developed to support sensors requiring only two such ASIC readout chips, for 64 total channels. This board provides signal conditioning for HERMES control and readout, as well as bias and temperature control. It connects via ribbon cable through the vacuum feedthrough to a standalone EPICS control unit which is addressable over Ethernet. The control unit also supplies high voltage for sensor bias and low voltage (current) for the thermoelectric coolers, and provides inputs for test pulses as well as analog output from a selectable channel for diagnostics.

3.4 System Test

The complete detector system pumps to a typical base pressure of 10^{-6} mbar. With a bath temperature of 12 °C, thermoelectric cooler current of 0.5 A provided a sensor temperature of -35 °C. Illumination was provided by NSLS beamline X16A, a test beamline outfitted to perform basic x-ray optics tests using a bending magnet source and Si(111) monochromator. The photon energy of the beam was 9.1 keV, collimated to less than 100 μrad and 10-20 μm beam size. The detector was mounted to a 4-axis positioner, with vacuum tee fitted to the conflat entrance port to provide vacuum pumping and gauging as well as use of a beryllium x-ray entrance window. For data acquisition, integration to spec (certif.com) was provided via a macro named hermes.mac which integrates data capturing and zeroing into the standard accumulation sequence.

Detector bias was set to 120 V based on electrical tests. Global threshold was optimized manually, yielding a noise count rate of 11 counts per 1000 seconds over all 64 channels (0.6 counts per hour per channel); since only 18 channels are expected to be used, the total noise count rate can be expected to be as low as 11 counts per hour (0.003 cps).

To characterize detector energy resolution, a multichannel analyzer was used to collect spectra from the analog output for typical sensor channels under synchrotron beam illumination of individual sensor strips. FWHM for low-rate, single photon pulses (~10 cps) was typically ~8% of the peak position (750 eV at 9.1 keV), as shown in figure 4.

4. Summary

A custom detector system, complete with cooling, vacuum enclosure, and controls was produced and tested within 10 months utilizing roughly 1 person-year of labor, by making efficient use of existing in-house designs and facilities. The demonstrated performance satisfies the requirements for NSLS-II IXS beamline, providing the required spatial resolution (80 μm pitch, 6 mm size), efficiency (98% efficiency for 400 μm silicon), low noise (0.003 cps), energy resolution ($\Delta E/E$ of 8%) and maximum count rate (100 kcps/ch over 18 channels for more than 1 Mcps total).

Future improvements to come include pixel-specific threshold trim automation and graphical data integration to the beamline control and acquisition system.

5. Acknowledgment

The authors wish to thank L. Reffi and D. Levy for detailed design and fabrication of the Tagma mechanical components, and D. Pinelli and D. Poshka for electronics assembly. Use of the National Synchrotron Light Source, Brookhaven National Laboratory, was supported by the U.S. Department of Energy, Office of Science, Office of Basic Energy Sciences, under Contract No. DEAC02-98CH10886.

6. References

[1] Cai Y Q, Coburn D S, Cunsolo A, Keister J W, Honnicke M G, et al. 2013 *J. Phys. Conf. Series* **425** 202001

[2] Shvyd'ko, Y V, Lerche M, Kuetgens U, Rüter, H D, Alatas A and Zhao J 2006 *Phys. Rev. Lett.* **97** 235502

[3] De Geronimo G, O'Connor P, Beuttenmuller R H, Li Z, Kuczewski A J, Siddons D P. 2003 *IEEE Trans. Nucl. Sci.* **50** (4) 885

[4] Honnicke M G, Keister J W , Conley R, Kaznatcheev K, Takacs P Z, et al. 2011 *J. Synch. Rad.* **18** 862

[5] Stetsko Y P, Keister J W , Coburn D S, Kodituwakku C N, Cunsolo A, Cai Y Q 2011 *Phys. Rev. Lett.* **107** 155503

Journal of Physics: Conference Series **493** (2014) 012016 doi:10.1088/1742-6596/493/1/012016

Scanning CCD Detector for X-ray Powder Diffraction

T Madden, J Baldwin, R Von Dreele, M. Suchomel, B H Toby

Advanced Photon Source, 9700 S. Cass Ave.
Argonne National Laboratory, Argonne, IL, 60439.

tmadden@aps.anl.gov

Abstract We discuss the design, fabrication and use of a custom CCD detector for x-ray powder diffraction measurements. The detector is mounted on a diffractometer arm, where line-by-line readout of the CCD is coupled to continuous motion of the arm. As the arm moves, the data from the CCD detector are accumulated and can be viewed as if it were a "film strip" with partial powder diffraction rings. Because of the unique design of the camera, both high-resolution and rapid measurements can be performed. Powder diffraction patterns are collected with speeds of a few minutes, or less, with many of the advantages of large area position-sensitive detectors, for example amorphous silicon flat panels, such as high sensitivity, direct evidence of grainy samples and freedom from low-angle asymmetry, but with resolution better than linear position-sensitive detectors and nearly as good as the ultimate in resolution, analyser-crystal detection [2,3].

1. Introduction

Powder diffraction is a simple form of X-ray diffraction performed on a collection of microcrystals, often in the form of a powder. Because it is advantageous if the crystals have random orientation and to have many crystals in the beam, a sample is often spun on a motor stage as it is exposed to x-rays to better randomize the crystal orientations and to improve particle sampling statistics. Because ideally all crystal orientations are represented, the diffracted x-rays from all Bragg peaks manifest as cones that project as concentric circles. Powder diffraction can be used to determine the structure of even complex materials, and is a vital crystallographic technique [1].

There are two types of powder diffraction instruments at Advanced Photon Source. At beamline 11-BM, a crystal analyser system, in which 12 photomultiplier tubes (PMT), each with an analyser crystal, are mounted on a goniometer arm; it offers the highest resolution powder diffraction data in the United States. The analyser-PMTs are moved on the goniometer arm to scan the projected powder diffraction rings [2]. Scans take typically one hour. For faster measurements at APS, several sectors use a stationary area detector such as a General Electric or Perkin-Elmer amorphous silicon flat panel detector, or less commonly a MAR image plate. In this setup, a single image is taken with a 0.1s to 30s exposure. This fast measurement is not nearly as high resolution as the crystal analyser system, but the fast measurement time allows for many otherwise impossible experiments [3]. It would be advantageous to build a detector that could take measurements with resolution comparable to the analyser system but speed closer to the area detector; such a detector, the Scanning CCD, is presented here.

2. Scanning CCD Detector

2.1. Synchronizing CCD Readout with Detector Movement

To understand how the Scanning CCD works, recall that an image is projected onto the active area of the CCD and incident photons are integrated as electronic charge in the CCD pixels. Once sufficient signal has been integrated over time, the image is read out by shifting it line by line. Each line is subsequently read out pixel-by-pixel to an output amplifier and digitized for storage in a computer. Generally the CCD is stationary and captures a stationary projection before being read out. In the Scanning CCD camera, the CCD is moved on a goniometer arm as it is read out. The shifting of the image line-by-line is exactly synchronized with the movement of the CCD. In this way, the image projected on to the CCD surface moves exactly with the image being read out. Furthermore, the CCD integrates the image while it is being readout. In a powder diffraction experiment, the CCD camera is moved on a goniometer arm through projected powder diffraction rings, and the camera returns a filmstrip, similar to that produced by a Debye-Sherrer Camera [4]. A diagram of a CCD moving up through a series of powder diffraction rings and returning a film strip is shown in Figure 1.

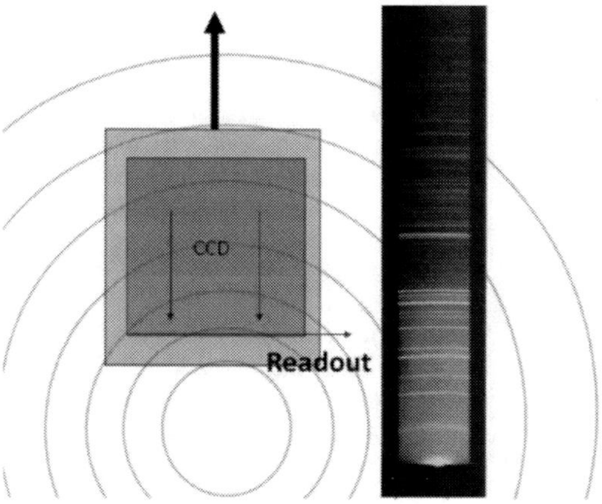

Figure 1. CCD moving though series of rings, generating a film strip. Readout of CCD is synchronized with camera movement.

2.2. Observations on Scanning CCD

One can observe several design considerations when building a Scanning CCD camera. The CCD must have very small pixels to preserve instrument resolution. Indeed, because the CCD is structured as discrete lines, CCD line shifting will slightly blur the image even if the device is smoothly moved on the goniometer arm. Obviously, a very accurate and repeatable motor system is needed. Because the motor system will move in steps unequal in size from the CCD line shifts, an algorithm must be designed to reconcile this difference. The slower the CCD moves, or the larger the CCD area, the larger the acquired signal. Because the CCD is moving, sensitivity variations in phosphor or optics in front of the CCD will be integrated over columns. To assure that image features are integrated over columns and not smeared across columns, the camera optics must have little or no spatial distortion and be accurately oriented with the direction of motion. Most commonly, data will be radially integrated once the filmstrip is recorded. The combination of radial and columnar integration means that every CCD pixel contributes to every point in the resulting 1-D powder diffraction pattern, rendering pixel-by-pixel sensitivity calibration unnecessary. Spatial distortion in the optics widens the peaks in the 1-D powder diffraction pattern.

2.3. Synchronization Algorithm

The synchronization algorithm accepts pulses from the motor system at the beam line. There will be M motor pulses corresponding to the spacing between each CCD line; the design requires that M be greater than or equal to 1, but allows this to be a real number. A digital accumulator integrates the number 1/M as each motor pulse is noted. Whenever the accumulator reaches 1.0, the integer portion of the accumulator is dropped, and a single CCD line is shifted out of the imager. In this way, the CCD readout is synchronized with the goniometer motor system, but a strict integer relationship between the angular motion and the CCD spacing is not required. A diagram of the algorithm is shown in Fig. 2.

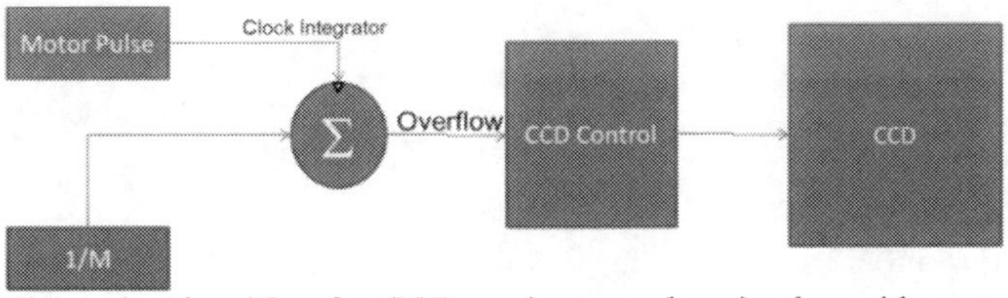

Figure 2. Algorithm for CCD readout synchronization with motor system. Motor pulses cause integration of 1/M in the accumulator; overflow causes a CCD line to be read out.

2.4. The APS Scan CCD Detector

The APS detector group designed and built a prototype Scanning CCD detector utilizing a Kodak KAF-16801 CCD with pixel dimensions of 9μm x 9μm coupled to a Scint-X structured scintillator via a fiber-optic faceplate. The electronics features all-digital double correlated sampling of the CCD signal and runs the motor synchronization algorithm on an on-board FPGA. Data is transferred from the camera to computer via Camera Link interface. The camera spatial resolution was measured to be 60μm FWHM. A photograph of the detector is shown in Figure 3.

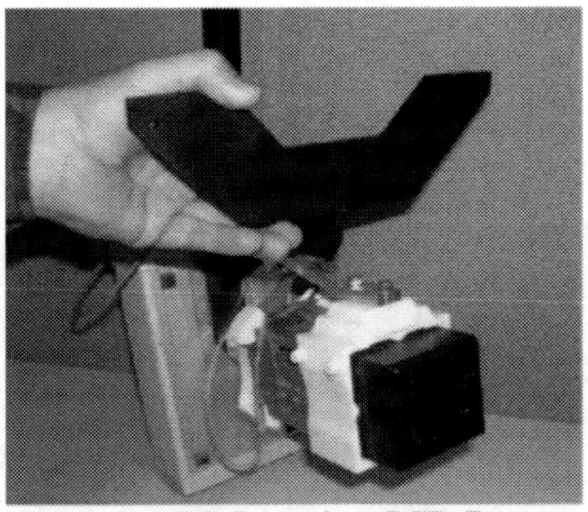

Figure 3. APS Scanning CCD Detector

3. Performance of the APS Scanning CCD Detector

A NIST standard (SRM 660a LaB_6) was measured at APS Sector 11-BM with four different detectors: the 11-BM crystal analyzer system, a Perkin Elmer Area Detector, a Mythen linear detector, and Scanning CCD [5]. To compare detectors, a measurement of resolution as $\Delta Q/Q$ was taken. The quantity Q is related to the diffraction angle, 2θ, and the crystal lattice spacing, d, by $Q=2\pi/d=(4\pi\sin\theta)/\lambda$. The smaller the $\Delta Q/Q$, the better the resolution of the powder diffraction measurement. Figure 4 shows plots of ΔQ/Q for the four detectors. As can be seen, the scanning CCD outperforms the Mythen and Perkin Elmer in resolution and is close to the 11-BM crystal analyzer system. These differences are not small, as can be seen in the comparison of widths for a single peak, shown in the same figure.

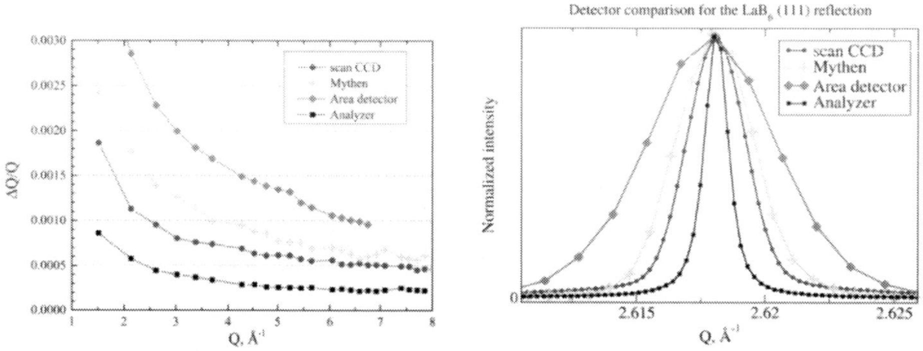

Figure 4. On left, comparison of ΔQ/Q for four detectors. On right, a single peak from all four detectors.

4. Conclusion and Acknowledgements

The APS Scanning CCD Detector provides resolution nearly of the quality of the best in U.S. 11-BM crystal analyser system, but can take data 10 times faster than the analyser system. With future improvements to the CCD detector, we expect to achieve a 20x improvement in speed over the 11-BM system. If multiple CCDs are employed, this speed gain can be further multiplied.

The authors would like to thank Linn Ribaud for assisting in data collection at APS and Dr. Pamela Whitfield of the National Research Council of Canada for providing a jadrarite sample. Use of the Advanced Photon Source, an Office of Science User Facility operated for the U.S. Department of Energy (DOE) Office of Science by Argonne National Laboratory, was supported by the U.S. DOE under Contract No. DE-AC02-06CH11357.

5. References

[1] Clearfield A, Reibenspies J, and Bhuvanesh N 2008 *Principles and Applications of Powder Diffraction* Hoboken, New Jersey: Wiley-Blackwell.

[2] Lee P L, Shu D, Ramanathan M, Preissner C, Wang J, Beno M A, Von Dreele R B, Ribaud L, Kurtz C, Antao S M, Jiao X and Toby B H 2008 *J. of Synchrotron Radiation* **15** 427-32

[3] Lee J H, Almer J, Aydiner C, Bernier J, Chapman K, Chupas P, Haeffner D, Kump K, Lee P L, Lienert U, Miceli A and Vera G 2007 *Nuclear Instruments & Methods in Physics Research Section a-Accelerators Spectrometers Detectors and Associated Equipment* **582** 182-4

[4] Debye P and Scherrer P 1916 *Physikalische Zeitschrift* **17** 277

[5] Patterson B D, Brönnimann C, Maden D, Gozzo F, Groso A, Schmitt B, Stampanoni M and Willmott P R 2005. *Nuc.Inst. and Meth. in Phys. Research Sect. B: Beam Interactions with Materials and Atoms* **238** 224-8

Novel High Efficiency Microcolumnar LuI_3:Ce for Hard X-ray Imaging

Z Marton[1*], Vivek V. Nagarkar[1], Stuart R. Miller[1], Charles Brecher[1], Harish B. Bhandari[1], Peter Kenesei[2], Stephen K. Ross[2], Jonathan D. Almer[2], and Bipin Singh[1]
[1]Radiation Monitoring Devices, Inc., Watertown, MA 02472, USA
[2]X-ray Science Division, Advanced Photon Source, Argonne National Laboratory, 9700 South Cass Avenue, Argonne, IL 60439, USA

E-mail: ZMarton@rmdinc.com

Abstract. We have developed a structured scintillator using a vacuum deposition technique that is suitable for manufacturing large area scintillators in a microcolumnar form. While providing high absorption efficiency, it also allows great temporal and spatial resolution X-ray imaging. Microcolumnar films of extremely fast and bright cerium-doped lutetium iodide (LuI_3:Ce) scintillator were synthesized. It has high density (~5.6 g/cm^3), high effective atomic number (59.7), bright green emission (540 nm range, well matched to commercial optics and CCD sensors), light yield exceeding 115,000 ph/MeV, and rapid, afterglow-free decay (~28 ns). This new scintillator could resolve the 153 ns bunch structure of the Advanced Photon Source (APS) at Argonne National Laboratory (ANL). Due to the fast, afterglow-free decay, and high efficiency of LuI_3:Ce, during the experiments performed at the 1-ID hard X-ray beamline at the APS, single 65 keV X-ray photons could be resolved with high signal-to-noise ratio and with temporal resolution better than 20ns. In the future, it will enable a wide range of hard X-ray (20 keV to 100 keV) imaging and/or high frame-rate applications such as dynamic studies of the structural and electrochemical properties of batteries using microtomographic X-ray imaging, internal corrosion in fuel cells, and time-resolved muscle diffraction experiments.

1.Introduction

Recent developments in synchrotron radiation sources have generated an urgent need for high-performance X-ray detectors that offer sub-microsecond to millisecond temporal resolution [1]. The dramatic increases in the brightness of such sources, including ultra-high photon fluence, in conjunction with beam divergence approaching nanometers [2], have made it possible now to conduct experiments that have heretofore been impractical or impossible. However, exploitation of the capabilities of these new sources requires a complementary enhancement of detector performance for hard X-rays, up to 100 keV. This energy is needed to penetrate many state-of-the-art engineering materials such as titanium alloys, aluminum-lithium alloys, and steels, to analyze chemical dynamics in novel devices, batteries and energy storage concepts. Such X-rays are used to investigate labile biological materials as well that withstand higher energies better than the 8-12 keV of conventionally

* Zsolt Marton, 44 Hunt Street, Watertown, MA, USA, Phone: +1-617-668-6934

used X-rays. Most synchrotron X-ray imaging detectors are based on CCD optical detectors coupled to scintillating phosphor converter screens [3, 4]. The familiar difficulties of slow readout speed and limited imaging area with all such traditional detectors have been addressed by various manufacturers by combining modular designs based on large-area CCDs or scientific-grade CMOS (s-CMOS) sensors with parallel high-speed output architecture sensors bonded to fiberoptic tapers, making it possible to design high frame-rate detectors [5]. Unfortunately, however, CCD/CMOS-based area detectors currently in use have been designed for soft X-rays, and they function poorly when used with hard X-rays.
We developed a process synthesizing structured thin films of a particular interest, cerium-doped lutetium iodide (LuI_3:Ce), invented recently at RMD [6, 7, 8], which is well-suited for manufacturing large-area films needed for demanding hard X-ray imaging applications using vacuum deposition technology.

2.Material and method

LuI_3:Ce, offers high potential for hard X-ray imaging applications due to its excellent properties. It has one of the highest conversion efficiencies (115,000 photons/MeV), a rapid initial decay (28 ns), an emission in the visible range (540 nm) [7, 8] with virtually no afterglow and can be readily fabricated into microcolumnar films for high resolution imaging applications [9, 10].
We performed co-evaporation runs using separate sources for the host and activator. However, we found that better control of the stoichiometry could be achieved by using doped targets of the desired composition rather than separate targets for dopant and host and trying to balance the respective vapor pressures. This change also substantially simplified the deposition process. We used our hot-wall evaporation concept to synthesize the films. It consisted of a silica crucible placed in a cylinder of wrap-around tantalum heaters. The disc-shaped substrate was placed on the top of the crucible and heated throughout the deposition process by a high-power quartz lamp to 350°C - 500°C. The entire setup was mounted in a vacuum chamber with a typical pressure in the $\sim 10^{-6}$ Torr range.

3.Results

3.1. Morphology

Our samples were characterized in terms of their morphology using SEM microscopy. The columnarity is evident in **Figure 1**. The images clearly show the anticipated columnar nature of resulting films, which is crucial for channelling scintillation light to provide the desired high spatial resolution in spite of their thickness. As the data presented below attest, even our 1 mm thick films demonstrated high spatial resolution of 250 µm, confirming our initial premise that the columnar nature of LuI_3:Ce films will mitigate the traditional trade-off between spatial resolution and efficiency, which is crucial for hard X-ray imaging.

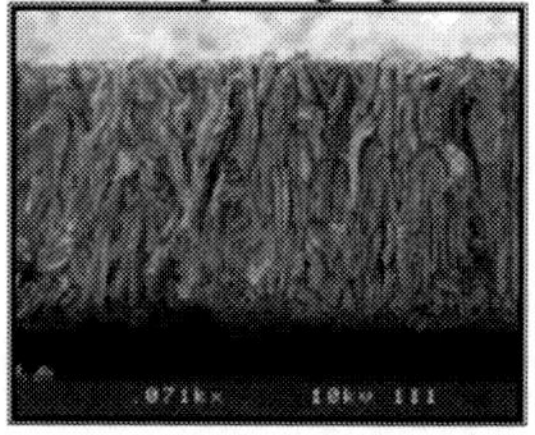

Figure 1 (Left) Microcolumnar LuI_3:Ce film showing the desired columnar structure. (Right) Side view of the same film showing tightly spaced, vertically oriented columns. This columnar structure is responsible to provide excellent spatial resolution in spite of its 1 mm thickness.

3.2 Scintillation properties

We measured afterglow on LuI_3:Ce and compared to our thallium-doped CsI standard. As **Figure 2** shows, our novel scintillator film displays virtually no afterglow next to CsI:Tl. The X-ray excited emission spectra (**Figure 2**, inset) of the LuI_3:Ce films were measured using the Cu-target X-ray generator (8 keV Cu Kα line) available at RMD. For the required flux at the sample, the X-ray generator was operated at 40 kVp with 20 mA current. The resulting scintillation light was passed through a McPherson model 234/302-0.2m monochromator and detected using an RCA model C31034 photomultiplier tube (PMT).
The typical spectroscopic shape of the scintillation emission from LuI_3:Ce crystal was known from work on single crystals before this film program had even begun [6]. The emission, centred at about

540 nm, is broad, but contains at least two incompletely resolved but unquestionably separate components. Although the intensities of these components differ somewhat, the correspondence between their locations demonstrates that our vapor deposition approach allows film growth with stoichiometry similar to that of the crystal.

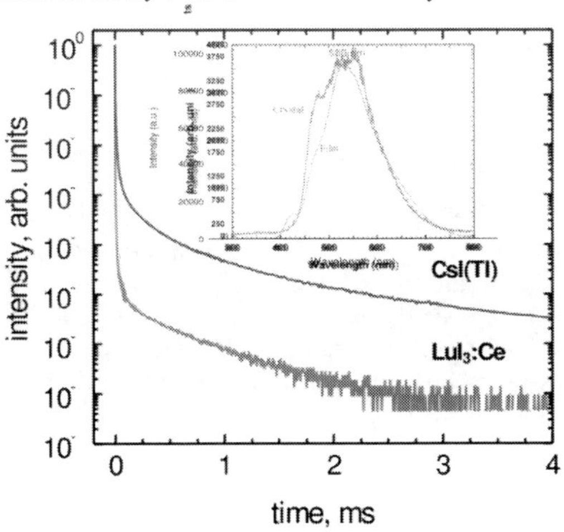

Figure 5. Afterglow characteristics of a LuI_3:Ce crystal with 0.5% Ce^{3+}. Also shown is an afterglow plot for a CsI(Tl) crystal under identical measurement conditions.

Figure 2 Semi-log plot of the scintillation decay of LuI_3:Ce (with 5% Ce) and standard CsI:Tl on exposure to a 50 ns X-ray pulse. The measured intensity at ~3 ms after the X-ray exposure has been stopped dropped to ~10^{-7} of the original value (when the X-rays were on). Thus, the afterglow in the LuI_3:Ce crystal is ~0.00001%, and is about two orders of magnitude lower than the standard CsI:Tl scintillator. In the inset, mission spectra for a representative LuI_3:Ce film under X-ray excitation. The corresponding data for a LuI_3 crystal [6] containing 5 mol% Ce confirms that the vapor deposition method can produce scintillation-grade LuI_3:Ce and that its emission properties may be adjusted by controlling growth conditions.

3.3 Spatial and temporal resolution

To obtain a quantitative measure of the spatial resolution of a 350 μm thick film, we calculated its modulation transfer function while exposing it to 70kVp X-ray. As it can be seen, this 350 μm thick LuI_3:Ce film has a resolution of 5 lp/mm (100 μm) with 10% modulation (**Figure 3a**). The screen certainly resolves higher line pairs, albeit with reduced contrast. It is important to note that the CCD camera used for these measurements has Nyquist limiting frequency of only 8.6 lp/mm corresponding to its 57 μm effective pixel size. Thus, one would expect to see even better resolution with a higher resolution camera.

The hard X-ray imaging and temporal resolution studies were performed at beamline I-ID of APS at Argonne National Laboratory. For imaging, we used an axially symmetric tungsten pin with flat top measuring ~100 μm in size. The pin was set in front of the scintillator and the images were acquired at strategically selected energies: 62 keV (just below the Lu K-edge); 65 keV (just above the edge); and 82 and 100 keV (considerably higher energies). These data are shown in the inset of **Figure 3**.

For temporal response measurements the film container was placed in the beam so that the X-rays would impinge on the LuI_3:Ce sample through the aluminum base and packaging, and the resulting light was measured on the opposite side by an APD. Two X-ray settings were used during these measurements, 82 keV and 65 keV. The APD output was coupled to a fast digital oscilloscope that stored the data. A representative oscilloscope trace is shown in **Figure 3b**, and, on an expanded scale, in the bottom image as well. The data shows two 65 keV APS pulses 153 ns apart, which correspond to the bunch structure of the APS. We have confirmed that each of these pulses is a signal from single photon interaction in the scintillator. These data demonstrate that the decay time of our scintillator is <20 ns.

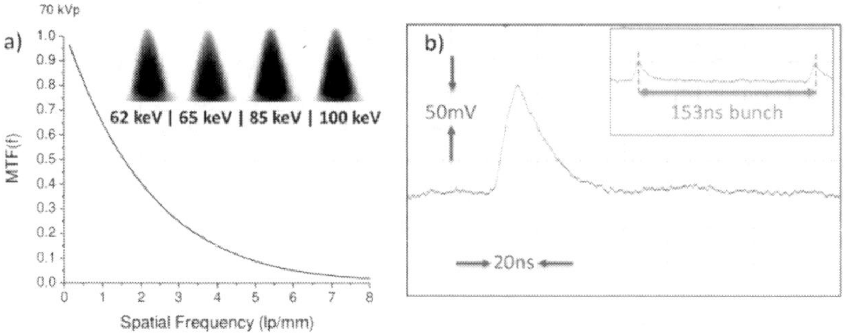

Figure 3 Modulation Transfer Function (MTF) of a 350 μm thick LuI_3:Ce film (a). The inset shows that the 1 mm thick LuI_3:Ce scintillator was able to resolve the 100 μm tungsten pin target. Images taken at energies below and above the Lu K-edge show no difference, indicating efficient and short range absorption of the fluorescence photons. Oscilloscope trace (b) shows fast decay time <20 ns of the LuI_3:Ce film. The inset data show two 65 keV APS pulses 153 ns apart, which corresponds to the bunch structure of the APS. Each pulse is a single X-ray photon, demonstrating the sensitivity and brightness of the scintillator, which permits detection of a single event.

4.Conclusion

While a variety of X-ray detectors satisfy the resolution, efficiency, and large active area needs for imaging with low-energy X-rays (8 keV to 12 keV), none is suitable for the hard X-rays (20 keV to 100 keV) and/or for high frame rate applications that are critical for numerous time resolved studies. We developed a process utilizing hot-wall evaporation that allows us to grow structured cerium doped lutetium iodide (LuI_3:Ce) scintillator, that will extend the applicability of current CCD/CMOS detectors to hard X-ray energies and will facilitate new detector designs with advanced features being sought to support such new studies. Its superior stopping-power, microcolumnar structure, fast decay enhance its brightness, spatial resolution and potential for timing resolution studies at high X-ray energies, respectively, while the low afterglow significantly reduces the required data acquisition time.

Acknowledgements

We thank Dr. Steve K. Ross and Dr. Peter Kenesei and Dr. Jonathan D. Almer at Argonne National Laboratory for carrying out hard X-ray imaging experiments at the 1-ID beamline of APS as well as for valuable discussions. This work was supported in part by the US Department of Energy under Grant DE-SC0007549. Use of the APS was supported by the US Department of Energy, Basic Energy Sciences, Office of Science, under Contract No. W-31-109-ENG-38. Use of the Advanced Photon Source, an Office of Science User Facility operated for the U.S. Department of Energy (DOE) Office of Science by Argonne National Laboratory, was supported by the U.S. DOE under Contract No. DE-AC02-06CH11357.

References

[1] 16th Pan-American Synchrotron Radiation Instrumentation Conference (SRI2010, Sept. 21-24, 2010), held at Argonne National Laboratory (Argonne, IL).

[2] National Synchrotron Light Source II (NSLS II), Brookhaven National Laboratory.

[3] Phillips WC, Stanton M., O'Mara D., Naday I. and Westbrook E. 1993 *SPIE* 1900

[4] Phillips WC, Stewart A, Stanton M, Naday I, Ingersoll C 2002 *J. Synchrotron Rad.* **9** 36

[5] Nagarkar VV, Singh B, Guo L, Gore D, Irving TC 2007 *Nuclear Instruments and Methods in Physics Research Section A* **576(1)** 38

[6] Shah KS, Glodo J, Klugerman M, Higgins W, Gupta T, Wong P, Moses WW, Derenzo SE, Weber MJ, and Dorenbos 2004 *IEEE Trans. Nucl. Sci.* **51(5)** 2302

[7] Glodo J, Shah KS, Klugerman M, Wong P, Higgins B, and Dorenbos P 2005 *Nuclear Instruments and Methods in Physics Research A* **537** 279

[8] Glodo J., van Loef, EVD., Higgins WH, and Shah KS 2008 *IEEE Trans. Nucl. Sci.* **55(3)** 1496

[9] Rodnyi PA, Physical Processes in Inorganic Scintillators (CRC Press, 1997)

[10] Knoll GF, Radiation Detection and Measurement, Second Edition, (John Wiley & Sons, 1989)

10 cm x 10 cm Single Gas Electron Multiplier (GEM) X-ray Fluorescence Detector for Dilute Elements

E H Shaban[1], D P Siddons[2] and D Seifu[3]

[1]Southern University, Electrical Engineering, Baton Rouge, LA 70813
[2]Brookhaven National Laboratory, NSLS, Upton, NY 11973,
[3]Morgan State University, Physics Department, Baltimore, MD 21251

Email: elhag_shaban@subr.edu

Abstract. We have built and tested a 10 cm x 10 cm single Gas Electron Multiplier (GEM) X-ray detector to probe dilute amounts of Fe in a prepared sample. The detector uses Argon/Carbon Dioxide (75/25) gas mixture flowing at a slow rate through a leak proof Plexi-glass enclosure held together by O-rings and screws. The Fluorescence X-ray emitted by the element under test is directed through a Mylar window into the drift region of the detector where abundant gas is flowing. The ionized electrons are separated, drifted into the high electric field of the GEM, and multiplied by impact ionization. The amplified negatively charged electrons are collected and further amplified by a Keithley amplifier to probe the absorption edge of the element under test using X-ray absorption spectroscopy technique. The results show that the GEM detector provided good results with less noise as compared with a Silicon drift detector (SDD).

1. Introduction

We used a 10 cm x 10 cm rectangular standard single Gas Electron Multiplier (GEM) to build a fluorescence detector [1-3]. Fluorescence X-ray is a characteristic secondary X-ray emitted from an element when it is bombarded by high energy X-ray beamline. Fluorescence X-ray experiments are used in X-ray fluorescence spectroscopy to analyze samples in ceramics, metals, geology and forensic science. There are a number of commercially available fluorescence detectors such as Lytle[4], Passivated Implanted Planar Silicon (PIPS) [5], and Silicon Drift Detector (SDD) detector [6]. In this poster we are developing another fluorescence detector using a GEM that was made and provided by 3M Corporation [7].

The intent is to use the GEM as an amplification medium of the primary electrons ionized by the low number of fluorescence X ray photons emitted from dilute elements in order to be able to identify and characterize the element. The low number of X-ray fluorescence photons ionize a detecting medium of gas that result in a very low number of the primary electrons. In order to detect the signal from the low count of the primary electrons an amplifier is required to amplify the signal above the background noise level present. In this experiment we used the GEM holes as a confined path for the electrons that are drifted by an electric field. Within the GEM holes path, a very high and controllable electric field is applied to accelerate the primary electrons in the presence of a detecting gas. The highly accelerated electrons by the strong electric field produce more electrons by impact ionizations with the neutral gas atoms. An avalanche of electrons is formed that provides a signal that can be easily detected, amplified using 428 Keithley current amplifier, and processed to provide a better high signal to noise (S/N) ratio absorption edge plot. This is a continuation to our previous experiment to build a fluorescence detector using a 3-inch diameter GEM that was donated by 3M corporation a few years ago [2]. In this experiment we have built a larger solid angle measuring 10 cm x 10 cm single GEM fluorescence detector. Figure 1 shows a schematic of the experiment. The detector is placed at right angle with respect to the test sample and the X ray beamline axis.

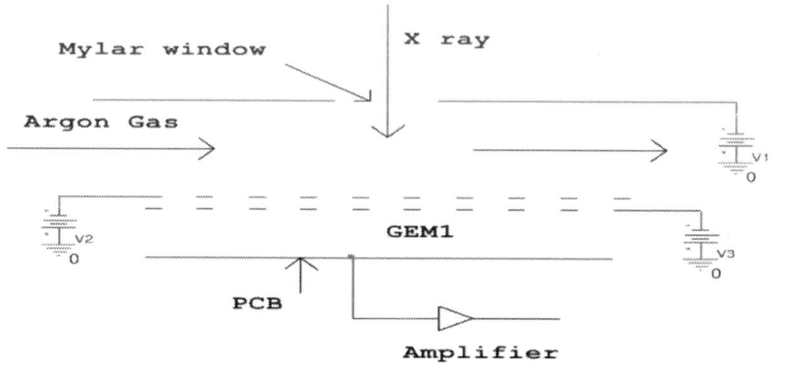

Figure 1: A sketch of a single GEM detector showing the Mylar window, the GEM and the PCB

2. Experiment

The GEM gas detector is formed from three regions known as the drift, amplification, and collection or induction regions. Figure 1 shows the single GEM gas detector that was built using Plexi glass machined in a rectangular shape measuring 19 cm x 19 cm with an open X-ray window measuring 10 cm x 10 cm, that has the same size as the GEM dimensions. The window entrance of the detector is covered by a one sided Aluminized Mylar placed a distance of about 2.5 cm away from the GEM upper conducting copper electrode. The GEM is separated from the collection region or the printed circuit board (PCB) using a Plexi glass flange about 0.3 cm thick. The bulk of the gas, Argon/Carbon dioxide (75/25) mixture, is allowed to enter and leave the drift region at a slow rate. The tightly enclosed detector allows the gas to fill the whole detector with the detecting gas with a controllable inlet and outlet passage located in the drift region. The drift electric field is applied between the Aluminized Mylar surface, termed as the cathode, and the upper copper conducting side of the GEM. The polarity of the applied voltage in this and subsequent regions is made such that the electron drift towards the GEM upper electrode and out of the GEM lower electrode towards the conducting copper coated PCB that acts as an anode and a virtual ground. In the drift region the ionized heavy charges are drifted towards the Mylar. A relatively very high electric field, as compared to that at the drift region electric field, is applied between the two copper electrodes of the GEM. This field can be changed to control the amplification of the primary ionized electrons or the gain of the GEM detector.

The printed circuit board (PCB) or the anode collects the abundant electron avalanche driven out form the GEM amplification region. The electric field magnitude in the three regions can be controlled by an adjustable High Voltage DC sources. The most critical electric field is within the GEM electrodes. It should be adjusted to avoid gas breakdown due to the spark formation that can damage the GEM. All parts of the detector are machined and held together using O-rings and screws to form a gas leak proof enclosure. The electron charge is collected from the PCB and fed to an external 428 Keithley current amplifier and then processed to display the absorption edge of the sample under test. We have used a two dimensional Maxwell Software to plot both the potential lines and the electric field lines in the three regions of the GEM detector. Figure 2 and 3 show the potential lines and the electric field strength in the GEM amplification region only. In this region the primary electrons drifting from the drift region enter the GEM holes and are amplified by impact ionization.

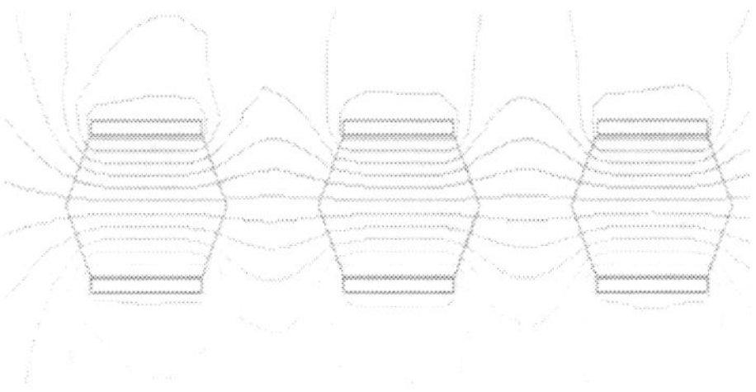

Figure 2: The equi-potential lines within the GEM and the Kapton Dielectric

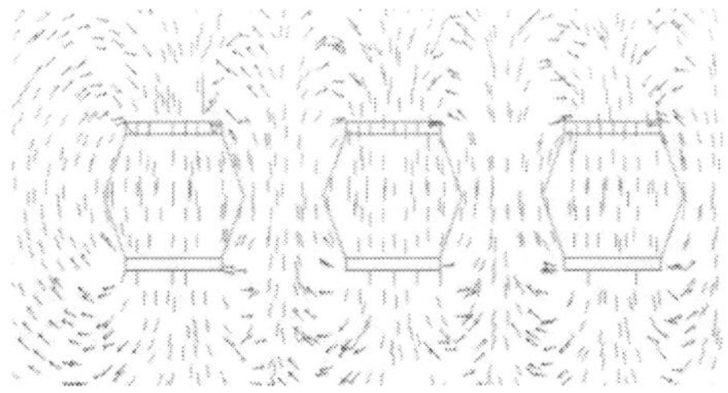

Figure 3: The electric field lines within the GEM

3. Results

A prepared dilute sample of Fe about (3-5%) is made at the Center for Advance Microstructures and Devices (CAMD) at Louisiana State University (LSU). The test sample was measured using the GEM detector and a Silicon Drift Detector (SDD) that is regularly used at the Double Crystal Monochromator (DCM) at CAMD. We found that the closer the GEM detector entrance to the sample the less noise scans we obtained. The single scans that we obtained improved dramatically from the one we first obtained when the detector was at a distance. The scan in Figure 4 is the best scan we obtained using the GEM detector. This is to demonstrate the ability of each detector to provide a single scan for comparison. Figure 5 is the same as Figure 4 but enlarge to compare the noise level in each detector.

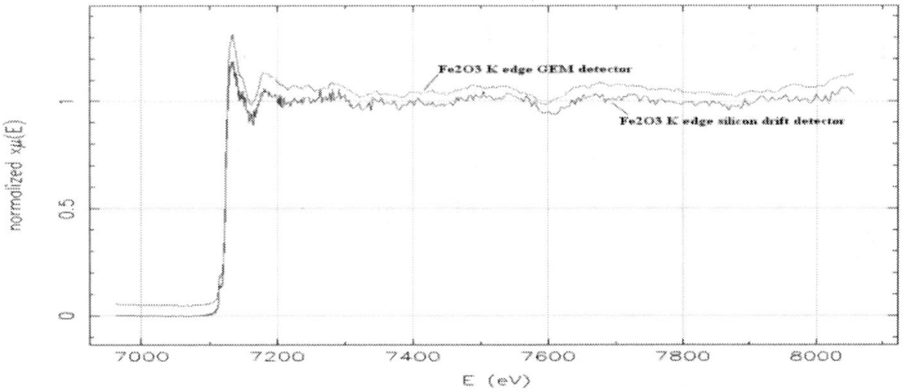

Figure 4: EXAFS scans for a dilute sample using the GM and SDD detectors

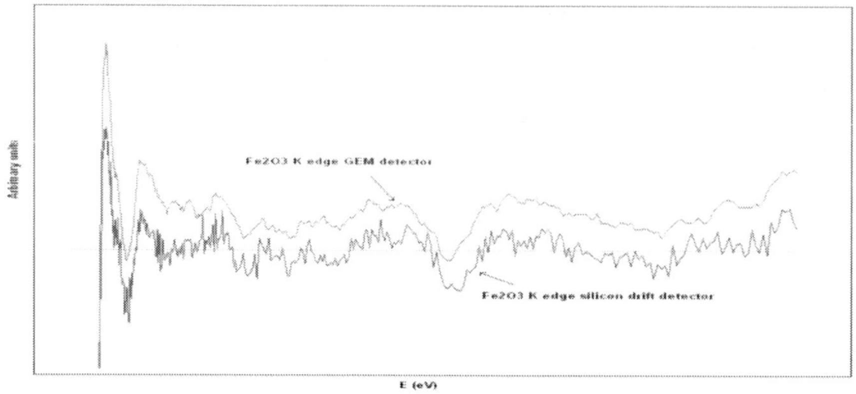

Figure 5: An enlarged diagram of figure 4 to compare the noise in the GEM & SDD detectors

Acknowledgement
This work is supported by NSF award DMR-1126444, and partly by LEQSF-EPS (2012)- PFUND-290. This work was done in collaboration with NSLS, the Instrumentation Department at Brookhaven National Lab, NSLS workshop, and Morgan State University. We thank the students Goings, Leonard, Driggs, Berthe, Bamagu, and Carriere for the help. We thank CAMD/LSU for providing the beamtime at DCM hutch, Dr. Roy, Merchant, and Jiles for the help & assistance, and 3M Corporation for providing the GEMs.

References
[1] Sauli F 1997 Nucl. Instr. And Meth. A 386 531
[2] Shaban E H, Siddons D P and Kuczewski A, 2007 Nucl. Instr. And Meth. A 582 (220), 186
[3] Bouclier R, Capeans A, Dominik W, Hoch M, Labbe J C, Million G, Ropelewski L, Sauli F and Sharma A 1997 IEEE Transaction on Nuclear Science, Vol. 44, No. 3, 646.
[4] www.ssrl.slac.stanford.edu/mes/xafs
[5] www.canberra.com/product/detector/pips-detector-technologoes.asp
[6] www.hitachi-hitec-science.us
[7] www.3m.com/wps/portal/3M/en-US/Austin/plant

A large area CMOS detector for shutterless collection of x-ray diffraction data

A C Thompson[1], E M Westbrook[1,3], W M Lavender[2] and J C Nix[3]

[1]Research Detectors Inc. 311 Wexford Ave, Lemont, IL, 60439, U.S.A.
[2]Illinois Institute of Technology, 3300 S Federal St., Chicago, IL, 60616, U.S.A.
[3]Molecular Biology Consortium, 835 S. Wolcott Ave., Chicago, IL, 60612, U.S.A.

E-mail: acthompsonx@gmail.com

Abstract. Recent developments in CMOS devices have improved their radiation hardness, response linearity, readout noise and thermal noise, making them suitable for x-ray crystallography detectors. Large (14.8 x 9.4 cm) CMOS sensors with a pixel size of 100 x100 microns are now available that can be butted together on three sides. We have fabricated a 6-tile system in a 2x3 array with a 28.2 x 29.5 cm continuous imaging area. To make an x-ray detector the CMOS sensor is covered with a 3 mm flat fibre-optic plate (for radiation protection) and a Gd_2O_2S:Tb scintillator screen. A special feature of these systems is that they can be read out continuously at 10 frames/sec with excellent dynamic range without interrupting data collection.

We have installed this system at beamline 4.4.2 of the Advanced Light Source synchrotron. Anomalous diffraction data were recorded without an x-ray shutter, rotating the crystal sample continuously with an exposure time of 0.1 sec/frame and a rotation speed of 1°/sec for 180 degrees. The 1,800 frame datasets were processed in D*TREK and XDS data analysis programs and experimental phases were determined in PHENIX. The crystallographic results are typically significantly better than equivalent data recorded on a conventional CCD system, due to the 10X finer angular resolution of the recorded data. Very large systems can now be made that would have an active area of 56 x 59 cm^2 with 33 x 10^6 pixels.

1. Introduction

Large area Complementary Metal Oxide Semiconductors (CMOS) detectors are now available with sufficient radiation tolerance and response linearity that they are now suitable for demanding scientific applications such as x-ray crystallography. The readout noise and readout speed have also been significantly improved in the last five years. Hasegawa et al[1] in 2009 used a single large CMOS device to demonstrate the advantages of these detectors for x-ray crystallography experiments. We have now developed a CMOS sensor with an active area of 9.4 x 14.8 cm^2 with 100 x 100 μm^2 pixels. The sensors are buttable on three sides so systems can be built with a large active area with no gaps between sensors. A special feature of these chips is that they can be read out continuously; while one pixel is being read, all the other pixels remain active. This feature allows x-ray diffraction data to be acquired in a shutterless mode, in which the crystal is rotated smoothly through 180° while 1800 image frames are recorded.

2. Description of Detector

We have developed a detector system suitable for use at synchrotron facilities. It is a 2x3 sensor array with an active area of 28.2 x 29.5 cm. As shown in Figure 1 the system has no gaps between sensors. The top of the sensor is covered with a 3 mm thick fiber-optic plate (to protect it from radiation damage), on top of which is placed a thin Gd_2O_2S:Tb phosphor plate that converts the incoming x-rays to optical photons, which are detected by the sensors. The sensors are cooled to below -5° C to reduce thermal leakage noise. Figure 2 shows how the system is packaged with the electronic processing cards below the detector with room to insert a cooled copper plate to cool the sensors. The direct coupling between the phosphor plate and the CMOS sensor gives high conversion efficiency to the system so that one 12 keV photon produces about 400 electrons in the chip. The dark accumulation rate is about 25 electrons/pixel/sec and the readout noise is about 180 electrons. The sensors are readout with 6 analog readout cards that each have five analog channels that use 14-bit digital to analog convertors that can be readout as fast as 30 Hz to a PC using a CameraLink interface. The complete system is mounted inside a 50 x 50 x 10 cm aluminum box weighing 32 kg with the analog processing cards mounted below the sensor and the digital interface card below the analog cards. The box has a 250 μm Be entrance window as shown in Figures 3. The detector is coupled to a fast PC with 64 Gbyte of memory and a 40 Tbyte fileserver. Another PC is coupled to the fileserver to enable rapid data analysis immediately after a data sequence is acquired. This allows data analysis of one data set to be done while another data set is acquired.

Figure 1. 2x3 array of six large-format CMOS sensors to cover a large imaging area.

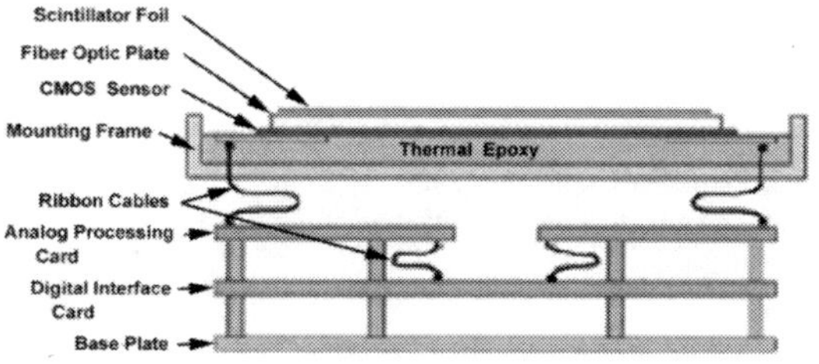

Figure 2. Arrangement of detector and electronic cards that permit cooling of the sensors.

3. Detector Performance

The commercial version of this detector was installed at the Molecular Biology Consortium beamline (4.2.2) of the Advanced Light Source. A brass plate with 1 mm holes spaced uniformly in a 5 mm grid was used to verify the minimal spatial distortion and a thin gold foil sample was used to calibrate the dose response and linearity of all pixels.[2] All pixels are corrected for their individual background count rate and gain before the images are written to disk.

The detector performance was measured by collecting diffraction data on a series of crystals from Olve Peerson of Colorado State University and Jack Tanner of the University of Missouri. Data were typically collected in shutterless mode over 180° with rotation speed of 0.2 - 1° per second and an image frame rate of 0.1 – 0.5 seconds per frame. Thus, 1800 images with an angular resolution of 0.1° per image can be collected in 3-15 minutes. Figure 4a is a single frame taken of a lysozyme crystal with an exposure time of 0.1 second and an angular range of 0.1°. Figure 4b is an enlarged part of a data frame taken of a crystal of histidine acid phosphatase (HAP), courtesy of Dr. Tanner, in which the spots, separated by only 0.8 mm, are clearly resolved.

Figure 3a. Front view of the detector.

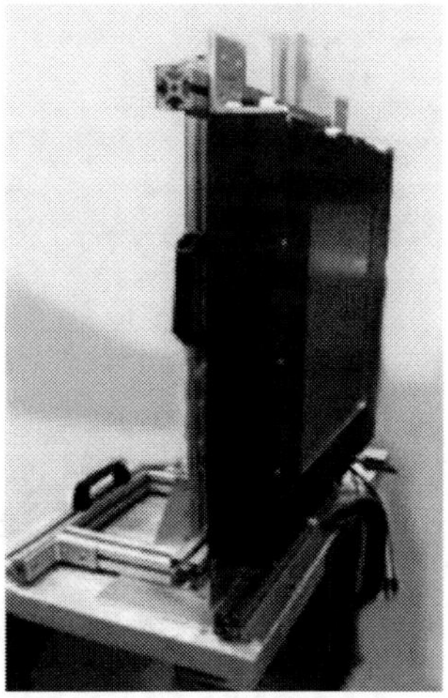

Figure 3b. Side View of the detector.

3. Data analysis

Dr. Tanner had substituted selenium for sulfur in the methionine residues of his HAP protein. Therefore diffraction from these crystals, stimulated by 12,700 eV x-ray radiation, exhibit an anomalous dispersion component to their Bragg reflections (the absorption edge of selenium is 12,658 eV). We collected 900 images, each of 0.2° and 0.2 sec/frame, for a total rotation range of 180°. The dataset was taken in 6 minutes at 12,700 eV with a detector-to-crystal distance of 150 mm. Data were processed both by Jim Pflugrath's dtprocess[3] and by Wolfgang Kabsch's XDS programs[4].

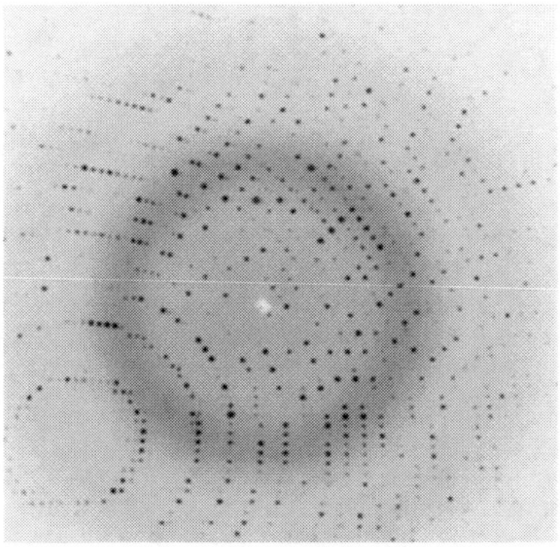

Figure 4a. Single frame of lysozyme taken in 0.1 sec, with 0.1° angular rotation of crystal.

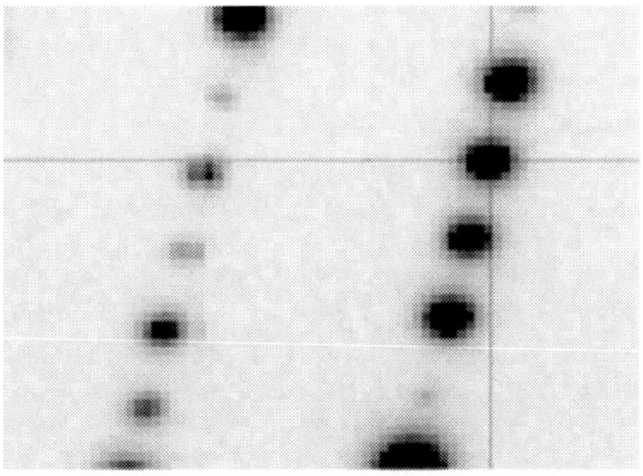

Figure 4b. Enlarged image of a HAP frame, showing peaks separated by only 0.8 mm.

The structure of histidine acid phosophatase (HAP) was solved with a single crystal. The HAP unit cell is 62 x 62 x 211 Å, $P4_12_12$. The dataset refined to a crystallographic R-factor of 18%, which is

Journal of Physics: Conference Series **493** (2014) 012019 doi:10.1088/1742-6596/493/1/012019

excellent for protein structures. The data processed with XDS to a resolution of 2.0 Å, as shown in the following table.

	RES	Rmerge	I/sig	completeness	redundancy	Anom R
	5.95A	3.2%	60.0	92.0%	6.95	5.87
	4.23A	2.9%	61.5	98.8%	7.66	4.14
	3.00A	3.7%	47.0	99.9%	7.76	3.04
	2.68A	5.2%	34.2	99.9%	7.78	2.40
	2.27A	9.1%	20.3	99.8%	7.76	1.57
	2.12A	12.2%	15.9	99.7%	7.73	1.36
	2.00A	18.3%	11.0	99.1%	7.68	1.15
Average		5.0%	30.8	99.3%	7.71	2.22

4. Summary

The new generation of large CMOS sensors with excellent sensitivity, linearity, radiation hardness and fast readout speed allow the fabrication of detector systems with improved performance compared to CCD systems. The ability to readout the detector while acquiring data reduces the time to take a complete dataset since the goniometer does not need to be oscillated as it does for CCD systems. In addition, taking more frames in a fine-phi slice mode produces higher quality data that can be refined to higher resolution than the standard 0.5° datasets used in standard CCD systems.

The CMOS technology we have developed now permits the fabrication of very large imaging x-ray detectors for macromolecular crystallography. We have recently developed a technique by which we can butt together our CMOS sensors on all four edges, with minimal restrictions. We can now make 12-tile, 18-tile, and 24-tile systems that can cover contiguous circle diameters respectively of 37.6 cm, 48.3 cm, and 56.4 cm as shown in figure 5. The 42.5 cm circular coverage of the Dectris Pilatus 6M silicon pixel array detector is also shown for comparison. Single sensor systems are also available.

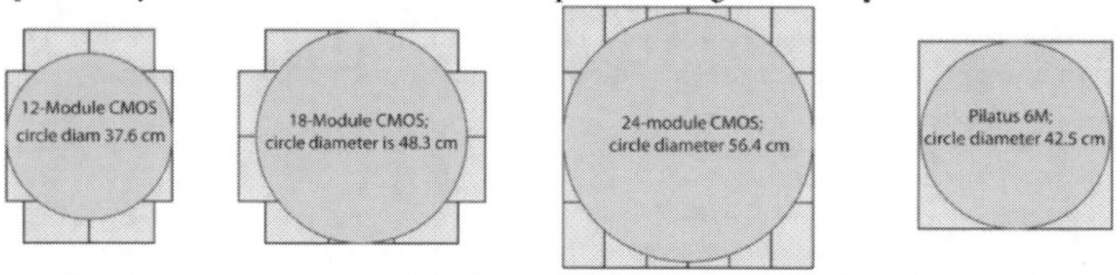

Figure 5. Diagram of three possible large CMOS detector configurations and the Pilatus 6M detector. The CMOS systems would have 100x100 μm^2 pixels and the Pilatus has 172x172 μm^2 pixels.

5. Acknowledgements

This work was supported by the U. S. National Institute of General Medical Sciences, through grants R43 GM085937-01 and R44 GM085937-02. The Advanced Light Source is supported by the Director, Office of Science, Office of Basic Energy Sciences, of the U.S. Department of Energy under Contract No. DE-AC02-05CH11231.

References

[1] Hasegawa K, et al 2009 Development of a shuterless continuous rotation method using an X-ray CMOS detector for protein crystallography, Journal of Applied Crystallography **42** (6) 1165.

[2] Thompson A, Nix J, Achterkirchen T and Westbrook E 2013 Large format CMOS-based detectors for diffraction studies, J. Phys.:Conf. Ser. **425** 012018.

[3] Pflugrath J, 1997 Diffraction-data processing for electronic detectors: theory and practice, *Meth. in Enzymol.* **276**, 286-306.

[4] Kabsch W, 1988 Evaluation of single-crystal X-ray diffraction data from a position-sensitive detector, *J. Appl. Cryst.* **21**, 916-924.

ALS-II, a Potential Soft X-ray, Diffraction Limited Upgrade of the Advanced Light Source*

H. Tarawneh[1], C. Steier, R. Falcone, D. Robin, H. Nishimura, C. Sun, W. Wan
Lawrence Berkeley National Laboratory, 1 Cyclotron Rd., Berkeley, CA 94720 USA

Abstract. The Advanced Light Source (ALS) at Berkeley Lab has seen many upgrades over the years, keeping it one of the brightest sources for soft x-rays worldwide. Recent developments in magnet technology and lattice design appear to open the door for very large further increases in brightness [1], particularly by reducing the horizontal emittance, even within the space constraints of the existing tunnel. Initial studies for possible lattices will be presented that could approach the soft x-ray diffraction limit around 2 keV in both planes within the ALS footprint.
Emerging scientific applications and experimental methods that would greatly benefit from ring based sources having much higher brightness and transverse coherence than present or near future storage ring facilities include nanometer imaging applications, X-ray correlation spectroscopy, diffraction microscopy, holography, ptychography, and resonant inelastic soft X-ray scattering at high resolution.

1. Introduction

Most designs under consideration for diffraction limited light sources make use of Multi-Bend Achromat (MBA) lattices. The first proposals for such lattices were made in the 90s and recently construction has started on the first implementation of the concept [2]. The required magnet strengths to realize small equilibrium emittances with those lattices are enabled by smaller vacuum chamber apertures and smaller magnet bores. This is possible because of recent advances in vacuum technology (NEG coating) magnet technology (wire EDM machining of poles) as well as advances in the understanding of nonlinear dynamics, beam-based calibration of lattices and the understanding of collective effects. The chosen candidate lattice for ALS-II is a Nine-Bend Achromat with a fully coupled beam and no damping wigglers. Figure 1 and 2 show the expected brightness performance of ALS-II and the coherent fraction for envisioned insertion devices.

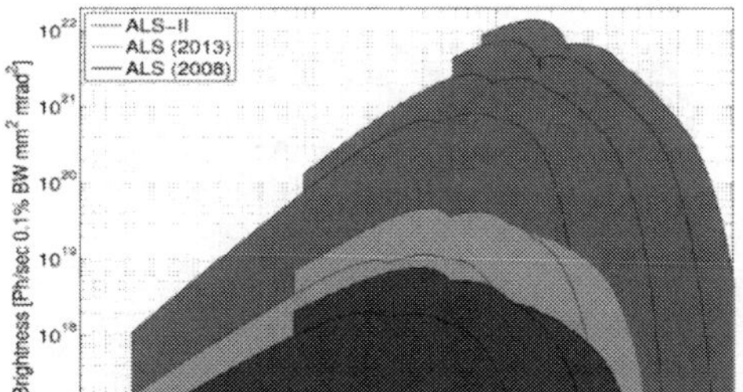

Fig.1: Brightness for ALS before upgrade (blue) after recently complete brightness upgrade (green) and for ALS-II lattice (red).

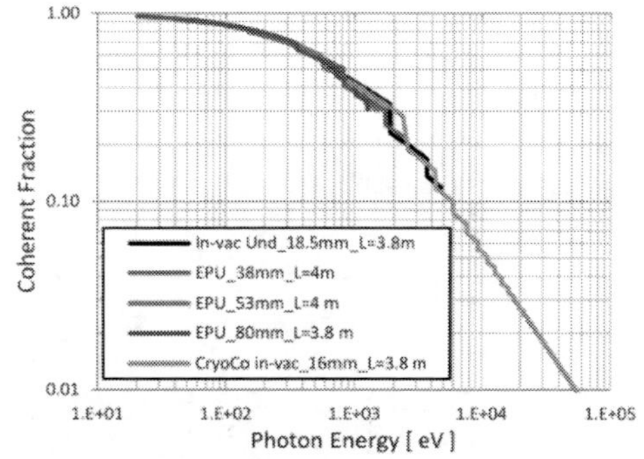

Fig. 2: Coherent fraction for ALS-II lattice for selected insertion devices.

2. ALS-II Scientific Case

The storage ring-based sources with diffraction-limited performance in both the horizontal and vertical plane will benefit many scientific applications of synchrotron X-ray beams e.g. nanometer imaging,

*The Advanced Light Source is supported by the Director, Office of Science, Office of Basic Energy Sciences, of the U.S. Department of Energy under Contract No. DE-AC02-05CH11231.
[1] hitarawneh@lbl.gov

spectroscopy, diffraction microscopy and soft X-ray scattering and result in an efficient use of the emitted photons [3, 4].
A 100x increase in coherent flux at ALS-II compared to the current ALS is driven mainly by three scientific application;

- Three dimensional imaging down to few nm resolution with chemical specificity, using techniques like ptychography.
- Q-resolved resonant inelastic x-ray scattering (q-RIXS) combined with dispersive spectroscopy making use of the full bandwidth of the undulator peaks and of the high spectroscopic resolution afforded by small source emittance in both planes.
- Correlation spectroscopy over various length (nm to μm) and time scales (ps to s).

3. ALS-II Candidate Lattice Overview

The proposed lattice for ALS-II is based on MBA, similar to what has been adopted at MAX-IV (under construction) and for upgrade plan at ESRF, PEP-X [5, 6]. The ALS-II lattice retains 12-fold symmetry and has a very compact arc structure facilitated by the fact that the magnet apertures, and hence the machine admittance, are defined by the small gap insertion devices. The small magnet aperture allows for the introduction of a strong defocusing gradient in the bending magnet and also a strong focusing gradient of the quadrupole magnets.
Initial optimization using multi-objective genetic algorithms has been carried out to improve the linear and basic non-linear lattice and yielded a candidate Nine-Bend Achromat lattice, Figure 3, with an emittance of 50 pm.rad at 2 GeV beam energy (fully coupled) and low beta function in the straight sections to optimize brightness and match the electron and photon phase space [7].
The proposed scheme for the chromaticity correction in the ALS-II lattice is based on compactness and flexibility. The small apertures in the sextupole magnets are used to introduce very strong sextupole fields. We avoided the introduction of these fields in the pole shape of the bending and quadrupole magnets where the flexibility of the lattice would have been compromised. The configuration of the matching cell which has relatively higher values of β_x, β_y and η_x functions is used to house four sextupole magnets families, two focusing and two defocusing ones, to correct the chromaticity.
The nonlinear effects driven by the strong sextupoles result in reduced dynamic aperture which hampers injection and reduces the Touschek lifetime. To overcome this, the ALS-II lattice will be operated at full coupling, i.e. with a round beam and an on-axis injection scheme. With on-axis injection electrons in already filled buckets in the storage ring are exchanged with newly injected buckets from an accumulator (swap-out injection [8]). The dynamic aperture and momentum aperture of the candidate lattice is sufficient even when introducing errors (> 200 $\sigma_{x,y}$) to allow high efficiency on-axis injection and provide decent beam lifetime (see Fig. 4).

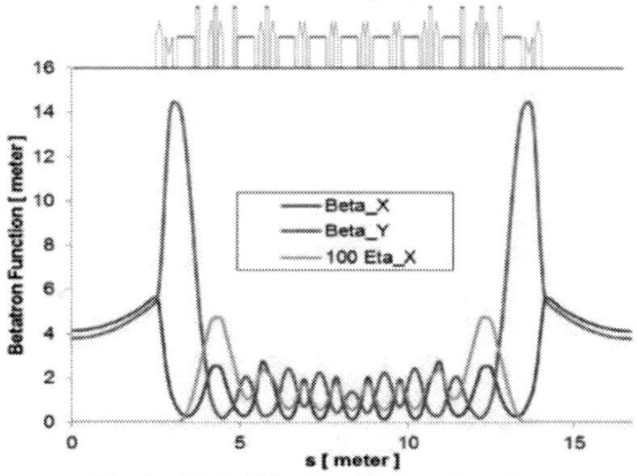

Fig.3: ALS-II lattice Twiss functions.

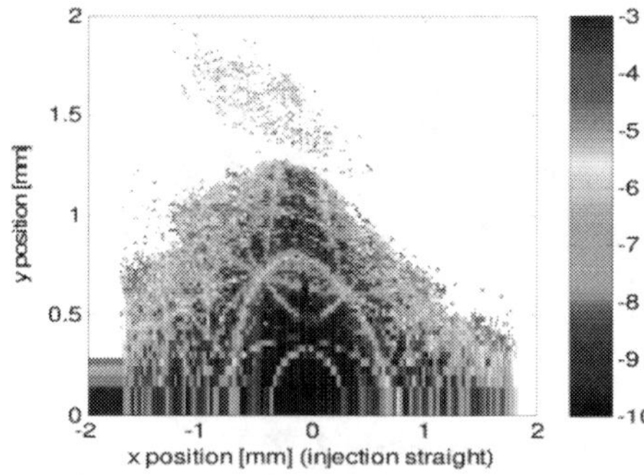

Fig.4: ALS-II Lattice dynamic aperture with errors

4. Intrabeam Scattering and Collective Effects

In electron storage rings, Intrabeam Scattering (IBS) can lead to an increase in the six dimensional emittance when the starting emittance is small, the bunch charge is high and the beam energy is moderate, as in the ALS-II case [9].

Mitigation of the impact of the IBS effect on the equilibrium emittance of the ALS-II lattice will be achieved by operating the lattice with full coupling where the vertical emittance is given entirely by betatron coupling and by bunch lengthening using a 3rd harmonic RF system. This leads to smaller electron density and hence less IBS growth rates in emittance, energy spread and bunch length.

Figure 5 shows an IBS growth rate comparison between a 500 MHz and a 100 MHz RF system with bunch lengthening based on the high energy approximation of the Bjorken-Mtingwa theory derived by K. Bane [9]. Given the small dispersion invariant H for the ALS-II lattice, the IBS growth rates are larger for the 100 MHz case even though the bunches are longer. This stems from the fact that the inverse dependence of the growth rates on the bunch length seems to be weaker than the dependence on the bunch population which is higher for the 100 MHz system with its fewer bunches.

As described above, the lattice uses small gap magnet elements to achieve low emittance. As a consequence, the resistive wall impedance becomes higher. In addition, small gap insertion devices also result in large resistive wall impedances, with apertures of 4 to 6 mm foreseen for the insertion devices. The bunch lengthening cavities, together with the choice of the fractional betatron tunes below the half-integer and the low momentum compaction factor, play an important role in mitigating this effect.

In the low emittance ring such as ALS-II ring single bunch instability thresholds can be relatively low due to the small momentum compaction factor and small synchrotron tune. Figure 6 shows a comparison of the single bunch current threshold due to the Transverse Mode Coupling Instability (TMCI) for a case with the natural bunch length and one case where the bunch is lengthened. Assumed is an impedance of 1 MΩ and Q-factor of 1 at 100 GHz resonant frequency. The benefit of bunch lengthening to increase the single bunch threshold is clearly visible [10].

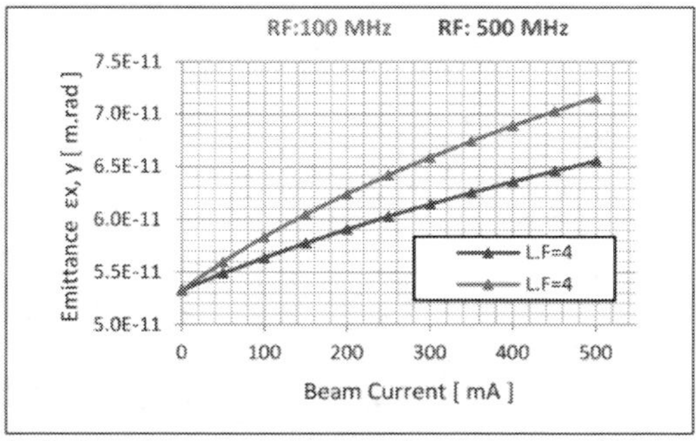

Fig.5: Steady state emittance dependence on beam current accounting for the effect of IBS.

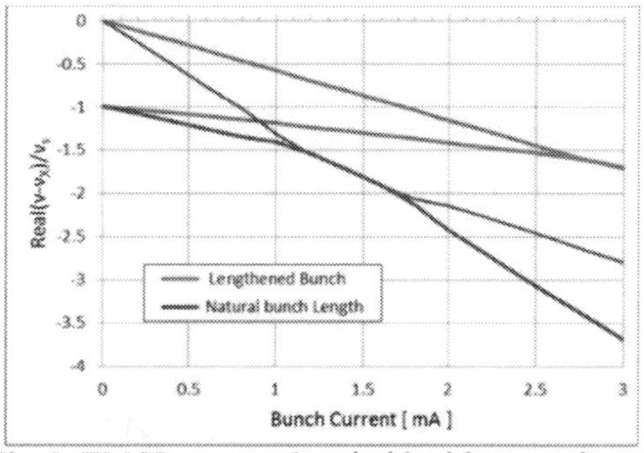

Fig.6: TMCI current threshold with natural as well as long bunch lengths.

5. Design Concepts for the ALS-II Magnets

The ALS-II lattice will be built around the insertion device, i.e. the ring admittance is defined by the low gap of the insertion devices and this gives the opportunity to use a small aperture within the magnetic elements of the arcs. This allows for the introduction of strong combined bending magnet and quadrupole and sextupole strengths. The compactness allows to achieve ultra low emittance with short circumference fulfilling the design goals and constraints. Another result is the small magnet size, which leads to less material to fabricate the magnets and low power consumption which is advantageous for cost considerations. Figures 7, 8 & 9 show conceptual designs of the ALS-II bending, quadrupole and sextupole magnets and Table 1 summarizes the maximum strength of the different types of the ALS-II magnets. The ALS-II lattice has no dedicated bending magnet instead uses shifted quadrupole magnets due to the high gradient required.

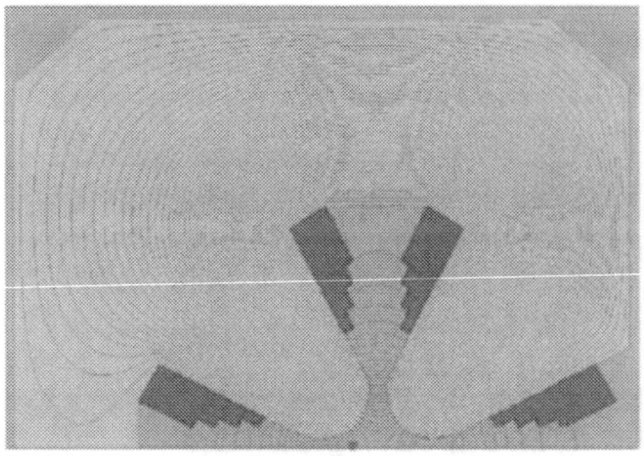

Fig.7: ALS-II Bending Magnet (shifted quadrupole magnet where beam axis marked with red dot).

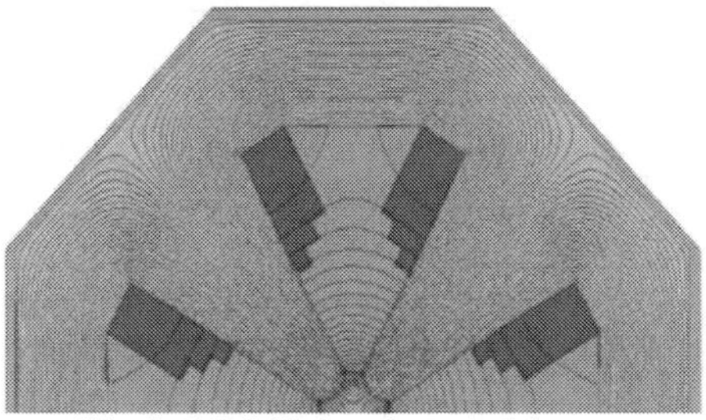

Fig.8: ALS-II Quadrupole Magnet.

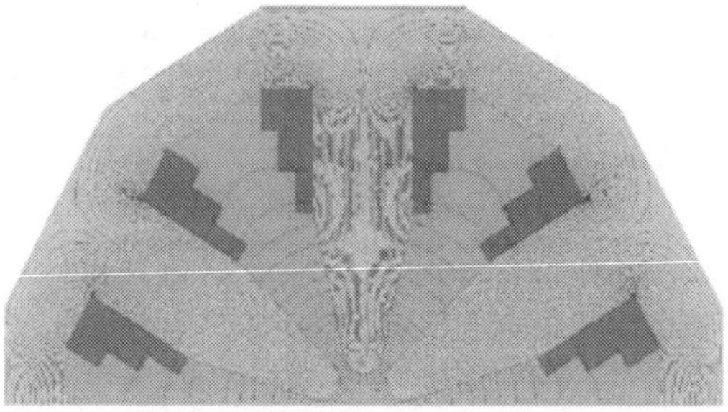

Fig.9: ALS-II Sextupole Magnet

Table 1. Maximum strength and bore radius of ALS-II lattice magnetic elements.

Type	Strength	Bore Radius
Unit Cell Dipole	0.77T / 50T/m	20 mm
Match. Cell dipole	0.77T / 20T/m	20 mm
Quadrupole	80 T/m	12 mm
Sextupole	3000 T/m^2	12 mm

6. Conclusion

The ALS-II candidate lattice has been designed based on the MBA concept that provides diffraction limited performance up to about 2 keV for a 2 GeV ring with slightly less than 200 m circumference, that fits in the existing ALS tunnel. The lattice has the same 12-fold symmetry as the existing ALS and retains the geometry of the existing straights. The dynamic aperture appears sufficient for decent Touschek lifetime and to support on-axis injection. Intra beam scattering appears manageable with 3rd harmonic cavities and bunch lengthening ratios similar to what is planned at MAX-IV. The magnet elements can be built from low carbon steel without suffering saturation effects that would produce large multipole errors. The predicted brightness of such a facility could exceed 10^{22} just above 1 keV photon energy and it would outperform any ring-based facility already in operation or currently under construction. Further design work is planned to solidify the concept and start detailed engineering design of key components.

7. References

[1] M. Bai et al., Nucl. Instr. and Meth. A (2010), doi:10.1016/j.nima.2010.01.045.
[2] Detailed Design Report on MAX IV Facility, First Edition, August 2010.
[3] ANL, BNL, LBNL, SLACNAL, Science & Technology of Future Light Sources, SLAC R-917, 2008.
[4] "Workshop on Science at the Hard X-ray Diffraction Limit", XDL 2011, Cornell, June 2011.
[5] A. Ropert et al., Proc. European Particle Accelerator Conference 2000, Vienna, 2000.
[6] R. Hettel et al., Proc. Particle Accelerator Conference 2009, Vancouver, 2009.
[7] C. Sun, et al., Phys. Rev. STAB 15, 054001 (2012).
[8] M. Borland, AIP Conference Proceedings 1234, 2009.
[9] K. Bane, SLAC-PUB-14892, 2012.
[10] Y.H. Chin, 'User's Guide for New MOSES', Version 2.0, CERN/LEP-TH/88-05.

NSLS-II Biomedical Beamlines for Macromolecular Crystallography, FMX and AMX, and for X-ray Scattering, LIX: Current Developments

M R Fuchs[1,4], R M Sweet[1], L E Berman[1], W A Hendrickson[1,2], O Chubar[1], N Canestrari[1,3], M Idir[1], L Yang[1] and D K Schneider[1]

[1] Brookhaven National Laboratory, Upton, NY, 11973, USA
[2] Columbia University and New York Structural Biology Center, New York, NY 10032, USA
[3] European Synchrotron Radiation Facility, 38043-Grenoble Cedex, France

E-mail: mfuchs@bnl.gov

Abstract. We present the current status of development of the two macromolecular crystallography (MX) beamlines, FMX and AMX, and the X-ray scattering beamline LIX, at the National Synchrotron Light Source-II (NSLS-II) [1]. Together, FMX and AMX will cover a broad range of use cases from serial crystallography on micron sized crystals, to very large unit cell complexes, to rapid sample screening, e.g. for the always-hard-to-grow membrane proteins and for ligand binding studies. The LIX beamline will support a variety of X-ray scattering measurements for studies on proteins in solution, lipid membranes and biological tissues. We have performed Synchrotron Radiation Workshop (SRW) [2] and Shadow[3] simulations to help select optimal methods to modify the size of the beam easily and smoothly at both FMX and AMX. The very low emittance of the NSLS-II storage ring and the resulting low divergence of the X-ray beam, as well as the long optical path lengths in the photon delivery systems lead to stringent requirements e.g. for vibrational stability and mirror quality. We discuss beamline design considerations addressing these challenges, such as combining mirror optics with compound refractive lenses (CRLs).

1. Introduction

Funded by the National Institutes of Health and the Department of Energy, the suite of Advanced Beamlines for Biological Investigations with X-rays (ABBIX) is scheduled to begin open user operation by 2016. The pair of MX beamlines is located at two identical canted undulators (IVU21) in sector 17-ID. The beamlines' specializations are complementary. The FMX beamline, for Frontier Microfocusing Macromolecular Crystallography, will deliver a high photon flux of 10^{13} ph/s at a wavelength of 1 Å into a spot of 1 µm width. It will cover a broad energy range of 5 – 30 keV, corresponding to wavelengths from 0.4 – 2.5 Å. Beam sizes up to 50 µm will be available. The AMX beamline, for Highly Automated Macromolecular Crystallography, will be optimized for high throughput applications, with beam sizes from 4 – 100 µm and an energy range of 5 – 18 keV (0.7 – 2.5 Å). The LIX beamline, for High Brightness X-ray Scattering for Life Sciences, will be accommodated in the neighboring sector 16-ID. Equipped with a single long undulator (IVU23), it will produce beams down to a size of ~1 µm via a two stage demagnification scheme, involving

[4] Author to whom correspondence should be addressed

mirrors and CRLs, and up to several hundred microns in size. It will have a broad energy range of 2.1 – 18 keV (0.7 – 5.9 Å) and be capable of simultaneously collecting data on three detectors.

2. Beam expansion

A variable beamsize is required to match the beam size to varying crystal sizes[4]. Additionally, for diffraction based crystal alignment schemes, rapid beam size increases to cover a large rastering area and locate crystals need to be interleaved with reliable refocusing for data acquisition on micron sized crystals. The secondary focusing stage of the FMX beamline will combine CRLs and Kirkpatrick-Baez (KB) focusing mirrors. For data acquisition, the curvature of the KB mirrors will be adjusted to move the focus downstream and match the beam size to the crystal size at the sample position. For rastering, by inserting CRLs upstream of the KB mirrors, the focus is moved upstream of the sample and thereby rapidly expands the beam size. By retracting the CRLs, the users can return within seconds to the optimized micro beam for crystallographic data acquisition.
In either focusing scheme, the figure errors of the mirrors will dominate both the minimally achievable beam size and the intensity variations on expanded beams. As beam distortions due to CRL inhomogeneities will be significantly smaller than distortions due mirror figure errors[5], all calculations have been performed assuming ideal lens shapes.

3. X-ray optics simulations

To derive specifications requirements for the mirror figure errors and avoid costly over-specifications, simulations were carried out both with the ray tracing package Shadow[3], and with the wavefront propagation package SRW[2]. To obtain a realistic estimate for the influence of mirror figure errors, both simulations were carried out using a measured (size) mirror file of the MISTRAL beamline at the ALBA Synchrotron (courtesy Josep Nicolas, ALBA, Spain). Its central 500 mm length section has a residual slope error (rms) $\delta = 0.1$ µrad and a height error (rms) of 7 nm (P-V = 21 nm). With this, the slope error limited vertical beamsize (fwhm) at an image distance of $q = 1.1$ m can be estimated to $2\times2.35\delta q = 0.5$ µm in good agreement with the simulated focus (figure 1, left). In SRW, the elliptical shapes of the KB mirrors were implemented using a grazing-incidence "thick optical element" propagator based on local ray-tracing. The KB surface height error was simulated by corresponding phase shifts ("masks") in the transverse plane at the mirror location.

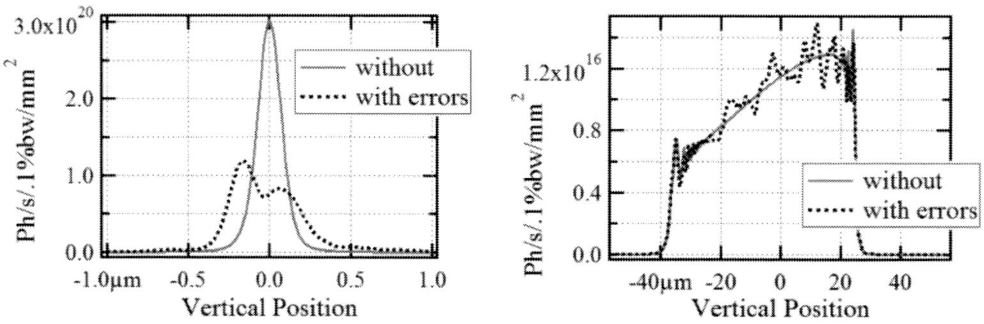

Figure 1. X-ray intensity distribution in vertical median plane calculated with SRW[2] at the sample position of the FMX beamline, for cases of best focusing (left) and defocusing using CRLs (right).

Due to the very small vertical emittance of the NSLS-II electron beam, the largest effect from the mirror figure errors are expected in the vertical direction. In the SRW-simulated beam profiles with and without mirror figure errors, significant modulations up to 20 % of the unperturbed profile can be observed, both for the focused beam spot at the sample position (figure 1, left) as well as for the beam defocused by insertion of CRLs upstream of the vertical KB mirror (figure 1, right).

To obtain a quantitative estimate of the transverse coherence, a "virtual Young's slit experiment" was simulated in SRW (figure 2) and the fringe visibility $F=(I_{max}-I_{min})/(I_{max}+I_{min})$ evaluated. The slit separation at half height is 270 μm for the vertical KB- and 22 μm for the horizontal KB-slits, providing a good estimate for the transverse coherence length at the FMX vertical KB mirror position.

Based on the van Cittert – Zernike theorem, for an extended incoherent source with a Gaussian density distribution in the transverse plane with different points emitting uncorrelated spherical waves, the fringe visibility after a Young's slit pair can be calculated analytically: $F=\exp(-2(\pi\sigma h/\lambda R)^2)$, where h is the slit separation, σ is the RMS source size, λ is the wavelength, and R is the distance to the source (figure 3). For the vertical slit separation plot, the values from the wavefront propagation agree very well with the calculation. For the horizontal case, the agreement is good with the exception of the higher fringe visibility in the SRW simulation at a separation of 100 μm, likely due to partial coherence at the secondary source, which in the analytical estimation is assumed to be incoherent.

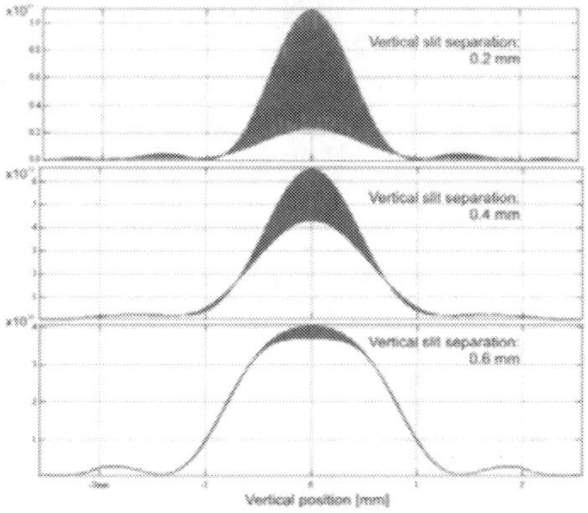

Figure 2. Intensity distributions in the vertical median plane 20 m downstream of a Young's slit pair at the position of the FMX vertical focusing KB mirror as a function of slit separation of 2 μm wide vertical slits.

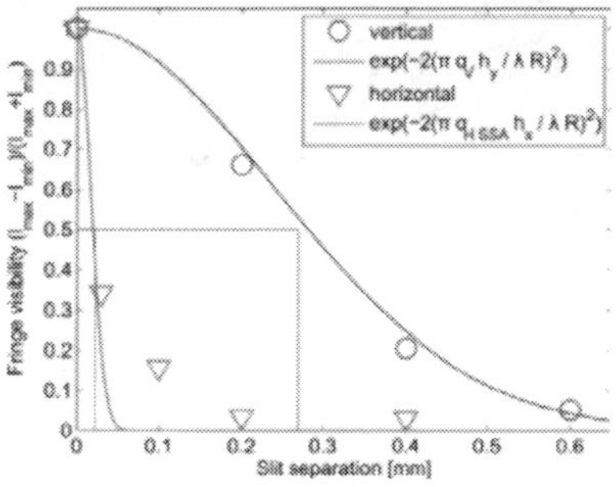

Figure 3. Fringe visibility with a slit pair at the position of the FMX vertical (blue) and horizontally (red) focusing KB mirrors as a function of slit separation of vertical and horizontal slits, respectively. The wavefront propagation was simulated with SRW [2].

These simulations provide clear evidence that even at the most critical mirror position the focusing is not dominated by coherence effects. However, with the vertical transverse coherence length being a significant fraction of approximately 1/6 of the total beam height, for the vertical KB mirrors for the AMX and FMX beamlines we used the Marechal criterion of the maximum tolerable wavefront error $<\lambda/14$ rms to specify a maximum rms height error. For an incidence angle $\theta = 2.5$ mrad at a wavelength of 1 Å, this limits the height error (rms) to < 1.5 nm.

4. Stability requirements of the double crystal monochromators (DCM)

The DCM with its central position in the photon delivery system is crucial to the beam stability. As both crystals have to be cooled with liquid nitrogen, they are subject to vibrational excitations from the cooling systems. To determine the maximally acceptable angular vibration amplitude for a DCM crystal, we assume a maximal vertical deviation of 10 % of the focal spot size of 500 nm at FMX. With a demagnification of 60, a deviation of 3 μm of the source position leads to this focus spot deviation. An angular deviation of a crystal to the downstream optics is indistinguishable from a source movement. A 2θ angular deviation corresponding to a 3 μm vertical source deviation is 75 nrad. For a single FMX DCM crystal and uncorrelated vibrations, the maximum deviation is 27 nrad. For AMX, an analogous estimate for the maximum 2θ angular deviation is 200 nrad.

The extreme angular stability requirement for FMX is a strong indication for a DCM geometry with a vertical axis, as this facilitates construction of a less vibration-sensitive design[6]. For AMX, both the less stringent angular stability requirement and the close proximity of the FMX white beam at the position of the AMX DCM indicate a design with a horizontal rotation axis.

5. LiX beamline layout

The optical layout of the LiX beamline has been revised since our last report[1]. The LiX monochromator will have two sets of interchangeable Si(111) crystals: a double crystal set to maintain a constant exiting beam height throughout the full energy range, and a channel-cut crystal for high stability used above 10keV. The first optical element is now a flat white beam mirror that deflects upward, which reduces the incident power onto the monochromator crystal, especially at low X-ray energies (~46 % at 3.6 keV). The vertical focusing mirror of the KB pair then deflects the beam downward by the same angle, so that the x-ray beam falls in the horizontal plane in the endstation.

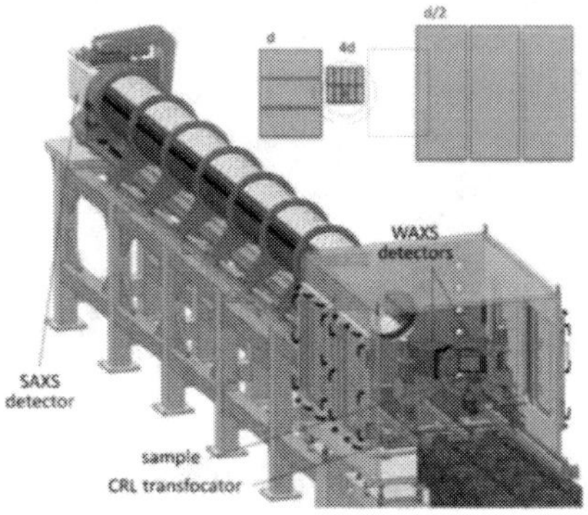

Figure 4. Layout of the LiX experimental station. Scattering data will be recorded by a Pilatus 1M SAXS detector (in-air) and two Pilatus 300K WAXS detectors (in vacuum). The inset (upper right) shows the solution scattering configuration for the 3 detectors, positioned at d/2, d and 4d from the sample to cover a wide, contiguous q-range. Both longitudinal and transverse positions of these detectors can be adjusted independently. The camera length for the SAXS detector is adjustable in the range of 2 – 5 m, utilizing a bellow-based flight path. The two WAXS detectors will be arranged in orthogonal orientations and can be positioned at 0.2 – 0.8 m from the sample.

The secondary source aperture is integrated into the same vacuum enclosure with several beam diagnostic components. The CRL transfocator is mounted on a long travel linear stage to provide variable demagnification and to compensate for the X-ray energy-dependence of the CRL focal length.

Conclusion

Critical details of the conceptual layout[1] for the FMX and AMX beamlines have been investigated. The optical and stability requirements due to the micron sized X-ray beam foci and the extremely low emittance of the NSLS-II electron beam were shown to be achievable with state of the art technology.

Acknowledgements

This work was supported by the U.S. National Institutes of Health, by the U.S. Department of Energy through contract DE-AC02-98CH10886 and by the New York Structural Biology Center.

References

[1] Schneider D K, Berman L E, Chubar O, Hendrickson W A, Hulbert S L, Lucas M, Sweet R M and Yang L 2013 *J Phys Conf Ser* **425** 012003
[2] Chubar O, Berman L, Chu Y S, Fluerasu A, Hulbert S, Idir M, Kaznatcheev K, Shapiro D, Shen Q and Baltser J 2011 *Proc SPIE* **8141** 814107–814107–10
[3] Sanchez del Rio M, Canestrari N, Jiang F and Cerrina F 2011 *J Synchrotron Rad* **18** 708–16
[4] Nave C 1999 *Acta Cryst* **D55** 1663–8
[5] Chubar O, Fluerasu A, Chu Y S, Berman L, Wiegart L, Lee W-K and Baltser J 2013 *J Phys Conf Ser* **425** 052028
[6] Paterson D, de Jonge M D, Howard D L, Lewis W, McKinlay J, Starritt A, Kusel M, Ryan C G, Kirkham R, Moorhead G and Siddons D P 2011 ed I McNulty, C Eyberger and B Lai *AIP Conf Proc* **1365** 219–22

The adaptive x-ray optic project at the Lawrence Livermore National Laboratory

T Pardini[1], L A Poyneer[1], T J McCarville[1], B Macintosh[1], B Bauman[1] and M J Pivovaroff[1]

[1]Lawrence Livermore National Laboratory, Livermore, CA 94550, USA

E-mail: pardini2@llnl.gov, Telephone: 1 925-423-1889

Abstract. Deformable mirrors (DMs) have been successfully used in astronomical adaptive optics at near-infrared wavelengths, greatly reducing atmospheric-induced aberrations [1]. Extending this capability to the soft and hard x-ray regime is now required in order to take full advantage of the beam quality characteristic of new generation synchrotron facilities and X-ray Free Electron Lasers (XFELs). Achieving this goal challenges both current mirror manufacturing techniques and physical optics modeling. The Lawrence Livermore National Laboratory (LLNL) is currently developing an x-ray DM to correct wavefront aberrations introduced along the beam path of a typical x-ray beam-line [2]. To model the expected performance of such a mirror, we have also developed simulation code based on the wavefront propagation library of functions PROPER [3]. Here we present the current status of the project, including metrology done on the mirror substrate. Additionally we report on results from our wavefront simulation code, which have proven very useful in predicting technical aspects of mirror deployment at a typical x-ray facility.

1. Why Adaptive X-ray Optic?

XFELs and modern 3rd generation synchrotrons are challenging our ability to handle their spatially coherent x-ray beams. While diffraction limited at the source, the beam accumulates phase and amplitude errors by interacting with transport x-ray mirrors during its propagation to the end-stations. The result is a diminished beam profile uniformity. Polishing x-ray optics to Ångström roughness and nanometer figure errors is mandatory to mitigate this problem; however the cost of these optics is considerable and their often harsh environment can contribute to degradation of surface quality with time. The demand for a highly uniform beam profile is common among FEL and synchrotron users; improved beam profile uniformity would increase resolution in imaging of biological samples; at the same time the capability of custom-shaping the beam profile would benefit experiments such as pump-and-probe (see fig 1). Adaptive X-ray Optic (AXO) promises to address these demands.

2. The Adaptive X-ray Optic Project at LLNL

LLNL is in the process of building a DM with the goal of quantifying its impact on wavefront propagation at a typical x-ray beamline. The AXO project is currently in its third year of funding. The project deliverables are (i) a 450mm long DM, (ii) a custom-built wavefront propagation code designed to predict mirror performance and optimize mirror integration with an x-ray beam line [2], and (iii) the development of a theory to analyze aliasing in wavefront

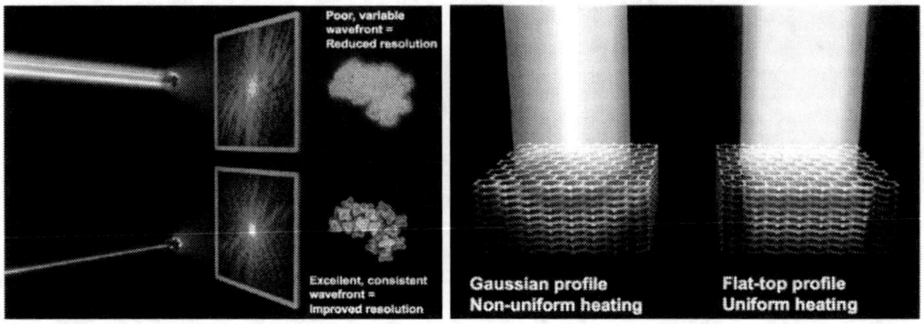

Figure 1. Artist rendering of AXO expected capabilities: (left panel) an aberration-corrected wavefront will enhance resolution in x-ray imaging experiments on biological samples. (right panel) The beam-shaping capability of AXO will allow uniform sample heating in ultra-fast transition experiments in solids. Rendering by Kwei-Yu Chu (LLNL)

sensing at x-ray wavelengths [4]. The project is the combination of LLNL extensive expertise in x-ray optics and adaptive optics applied to astronomy.

2.1. The hardware

A 450mm-long single crystal silicon substrate has been super-polished to 3.5Å rms roughness and 3.5nm rms figure error. The substrate is 40mm thick and 30mm wide; its polished surface has been extensively characterized at LLNL via long aperture interferometry, white-light interferometry and atomic force microscopy. Fig 2 shows a selection of data collected at LLNL. Deformable surface technology is provided by AOA Xinetics (Northrop Grumman); 45 1-cm wide actuators are being installed on the back of the mirror, with one strain gauge per actuator. Calibrated strain gauges will be used at first to provide close loop operations even without a wavefront sensor. The fully actuated mirror is expected at LLNL by August 2013.

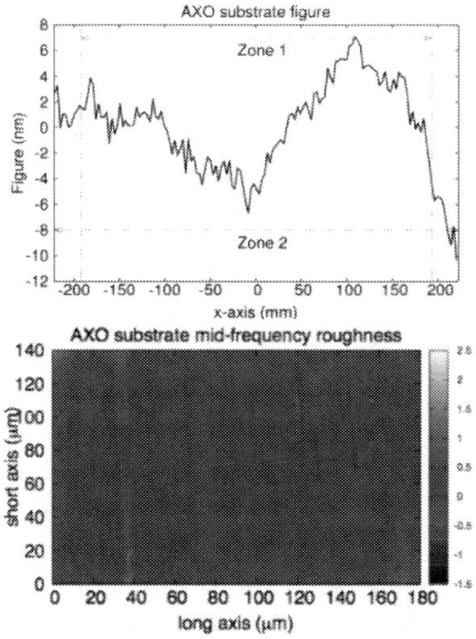

Figure 2. Selection of metrology data acquired on the AXO substrate: (Top panel) Large aperture interferometry data across mirror full aperture; (bottom panel) White light interferometry data taken in the center of the mirror clear aperture (x=0) with a 40x lens.

2.2. The software

Successful implementation of an x-ray DM strongly relies on the availability of a wavefront propagation code to predict mirror performance and help with deployment of the mirror at an x-ray facility. LLNL has developed a full physics optic propagation code based on the PROPER library of functions written for IDL [3]. In this manuscript we present simulations run using the beam transport system of the Linac Coherent Light Source (LCLS, SLAC) [5] as a test case. Photons with 8 keV energy and a gaussian beam with $60\mu m$ beam waist were used in all simulations. The goal of these preliminary simulations is to quantify the kind of improvement that will be achieved once we deploy our DM at a state-of-the-art x-ray facility. Figure 3 shows the simulated unfocused beam profile at the far experimental hall (FEH) of the LCLS after reflection off the two mirrors in the hard x-ray section of the beam transport system [6]. The low intensity wings at both sides of the peak are caused by figure errors on the offset mirrors. The ideal, diffraction limited beam is also shown for comparison. The result agrees with the published work of Barty and coworkers [7].

In order to estimate wavefront correctibility provided by our DM, we ran additional simulations replacing the two LCLS hard x-ray offset mirrors with two adaptive mirrors; these adaptive mirrors have the same specifications as the one under development at LLNL, including figure errors consistent with the results of our metrology on the mirror substrate. Code output confirmed that a considerable improvement in beam profile uniformity would be achieved within the actuation range of the mirrors. The result of this simulation is shown in Fig 4.

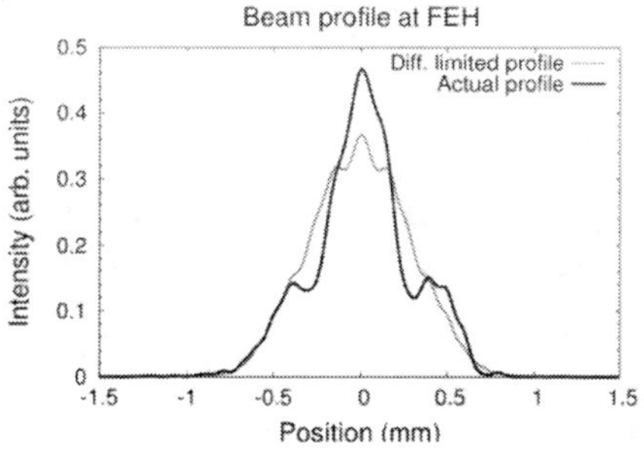

Figure 3. Beam profile at the Far Experimental Hall of the Linac Coherent Light Source (solid line). The diffraction limited beam profile is also shown as a red line.

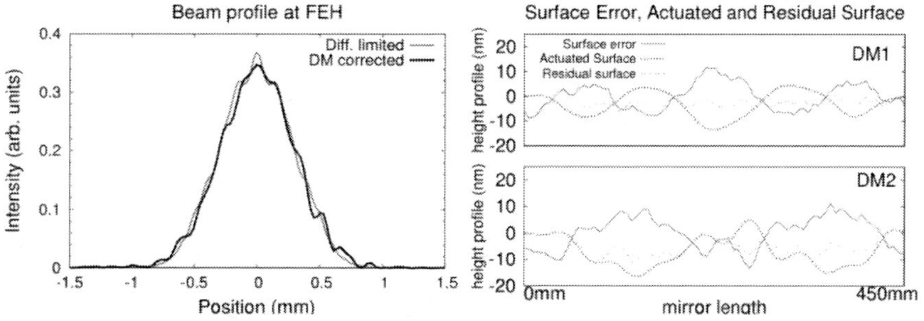

Figure 4. Left panel: beam profile at LCLS FEH hutch; diffraction limited and DM corrected profile are shown. right panel: Figure error, actuated surface and residual surface for the two DMs. Simulation was run using a 1536x1536 size grid

In a different mirror deployment scheme, a single DM could be used to correct phase errors introduced by upstream or downstream optics. Complications related to the Talbot effect must then be considered. A sinusoidal signal introduced in the wavefront will cycle between phase and amplitude error as a function of propagation distance [8]. A single DM can only correct phase errors; therefore deployment location of the mirror is critical to achieve optimal wavefront correction. In fig 5 we report the results of a simulation where a single DM was deployed downstream of the second hard x-ray offset mirror at the LCLS beam transport system. The beam profile is plotted as a function of DM location, which was varied between 2.5m and 7.5 m from the second offset mirror. The optimal DM deployment location is found to be 5m downstream of the second offset mirror. This shows how our code will be very useful in determining optimal mirror deployment location once the beamline parameters are known.

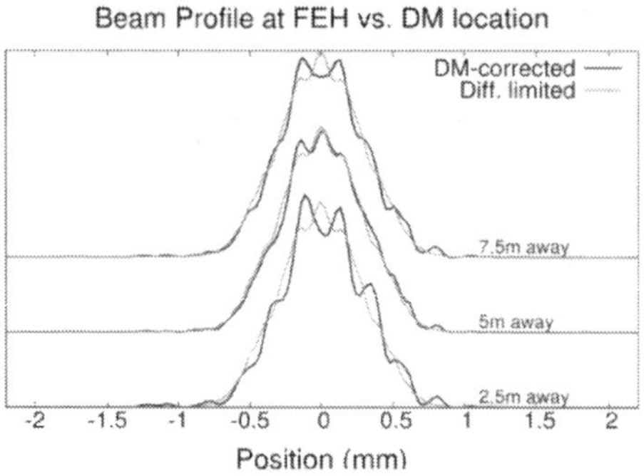

Figure 5. Diffraction limited and DM corrected beam profile at the FEH as a function of DM location with respect to the last LCLS offset mirrors. The dependence of the wavefront correction efficiency on the DM location is evident

3. Conclusions

LLNL, in collaboration with Northrop Grumman AOA Xinetics Inc., is currently developing a DM to be deployed at an x-ray light-source to achieve wavefront correction at x-ray wavelengths. Deployment of the mirror at an x-ray facility is expected by the beginning of calendar year 2014. As part of the project we are also developing code based on wavefront propagation necessary to predict performance of our DM. In this manuscript we have shown how our custom-built code will provide useful insights on DM performance and optimal deployment location depending on the chosen x-ray beamline parameters.

4. Acknowledgments

This work performed under the auspices of the U.S. Department of Energy by Lawrence Livermore National Laboratory under Contract DE-AC52-07NA27344. Document Release Number LLNL-PROC-646173

References

[1] Marois C, Macintosh B, Barman T, Zuckerman B, Song I, Patience J, Lafreniere D and Doyon R 2008 *Science* **322** 1348
[2] Pardini T, Poyneer L, Plinta A, Cavaco J and Pivovaroff M 2012 *Proc. SPIE* **8503** 85030H
[3] Krist J E 2007 *Optical Modeling and Performance Predictions III* **6675** 66750P
[4] Poyneer L, Baumann B and Macintosh B 2012 *Proc. SPIE* **8503** 85030I
[5] Emma P, Akre R, Arthur J, Bionta R, Bostedt C, Bozek J, Brachmann A, Bucksbaum P, Coffee R, Decker F J, Ding Y, Dowell D, Edstrom S, Fisher A, Frisch J, Gilevich S, Hastings J, Hays G, Hering P, Huang Z, Iverson R, Loos H, Messerschmidt M, Miahnahri A, Moeller S, Nuhn H D, Pile G, Ratner D, Rzepiela J, Schultz D, Smith T, Stefan P, Tompkins H, Turner J, Welch J, White W, Wu J, Yocky G and Galayda J 2010 *Nature Photonics* **4** 641
[6] McCarville T J, Stefan P M, Woods B, Bionta R M, Soufli R and Pivovaroff M J 2008 *Proc. SPIE* **7077** 70770E
[7] Barty A, Soufli R, McCarville T, Baker S L, Pivovaroff M J, Stefan P and Bionta R 2009 *Opt. Express* **17** 15508–15519
[8] Zhou P and Burge J H 2010 *Appl. Opt.* **49** 5351–5359

17th Pan-American Synchrotron Radiation Instrumentation Conference (SRI2013) IOP Publishing
Journal of Physics: Conference Series **493** (2014) 012023 doi:10.1088/1742-6596/493/1/012023

A Sagittally Bent Crystal for the Short Pulse X-ray Beamline at the Advanced Photon Source

Philip Strons, Ali Khounsary, Mikhail Antimonov, Eric M. Dufresne, Ruben Reininger

Argonne National Laboratory, 9700 S Cass Ave, Lemont, IL 60439, USA

E-mail: amk@aps.anl.gov

Abstract. The Short Pulse X-ray Imaging and Microscopy (SPXIM) Beamline at the Advanced Photon Source (APS) is being designed to provide a short-pulse focused beam at the sample some 60 m from the source. The second crystal in the horizontally diffracting double-crystal monochromator system is sagittally bent to focus the full vertical fan of the beam, preserving the flux of a typical undulator beamline at the APS, while providing a time-angle correlated beam on the sample with a few-picosecond pulse duration. An energy scan using this system requires changing the bent radius of the second crystal. Undesirable distortions in the sagittally bent second crystal are due to anticlastic deformations. In this paper, the analysis and design of the bent crystal is presented including an examination of the mechanical interaction between the crystal and the two pairs of rollers used to bend it. Ray tracings including the anticlastic deformations obtained from an optimized finite element model show a minimal effect on the spot size at the sample position.

1. Introduction

The proposed SPXIM beamline for the APS upgrade [1] uses a horizontally diffracting double-crystal monochromator (DCM). The thermal load on the first crystal of the DCM can be handled using existing cryogenically cooled designs. The sagittally bent second crystal (see Fig. 1) will collect most of the photons generated when the chirped electron beam, which has a correlation between vertical divergence and time, traverses the undulator. The bent crystal focuses the beam along the vertical direction at the timing slits, which are 54 m downstream from the source. To maximize the throughput of the bent crystal, the optimal design must minimize deviations from the desired cylindrical shape. The focusing at the sample position is done with a Kirkpatrick-Baez (KB) mirror system in which a pair of in-tandem mirrors is used. One mirror focuses the beam vertically and the other horizontally, to the same spot [2]. The first KB element is a horizontally deflecting elliptical mirror downstream of the timing slits that focuses the beam along the horizontal direction. A vertically deflecting elliptical mirror images the time slit along the vertical direction.

Undesirable distortions in sagittally bent crystals are due to anticlastic deformations. As a plate is bent in the sagittal direction, the non-zero Poisson's ratio of the material leads to bending in the tangential, or anticlastic, direction (see Fig. 2). Three common design approaches exist to decrease anticlastic bending: ribbed crystals [3], slotted crystals [4], and flat-plate geometries with an optimized aspect ratio of length to width [5].

Kushnir et al. [5] numerically determined the optimal aspect ratio for flat-plate crystals, assuming isotropic material. Their results revealed that the optimal aspect ratio for a sagittally

Journal of Physics: Conference Series **493** (2014) 012023 doi:10.1088/1742-6596/493/1/012023

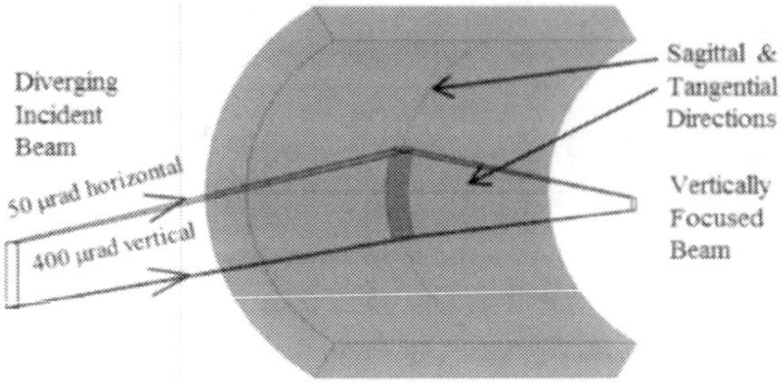

Figure 1. Schematic representation of the sagittally bent crystal for the SPXIM beamline.

Figure 2. Example of anticlastic bending in a bent crystal.

bent crystal depends only on Poisson's ratio and how the crystal is held and bent. There are analytical solutions for idealized plate and boundary conditions; however, they do not always accurately reflect the behavior of an actual bent crystal. Here, we report on the preliminary design of the sagittally bent crystal for the SPXIM beamline, determined numerically using a finite element (FE) model and including the effects of the roller/crystal interface.

2. Preliminary analysis of the anticlastic deformation

The approach to optimizing the crystal to minimize anticlastic distortion involves calculating the appropriate sagittal bending radius based on the beamline parameters, selecting the initial crystal plate geometry based on the beam footprint, analytically calculating the necessary moment to bend the crystal to the desired radius, converting the bending moment to an equivalent force, applying the force using a simplified FE model, and then refining the model. The roller/crystal interface is added to the FE model in section 3.

The crystal plate material used is silicon (111). Although silicon is not an isotropic material, the stiffness and Poisson's ratio are uniform in the (111) plane and thus in both the sagittal and tangential directions [5]. Idealized boundary conditions of simply supported edges and two lines of force were used to represent the crystal being bent by rollers (see Fig. 3). The crystal plate was assumed to have uniform thickness.

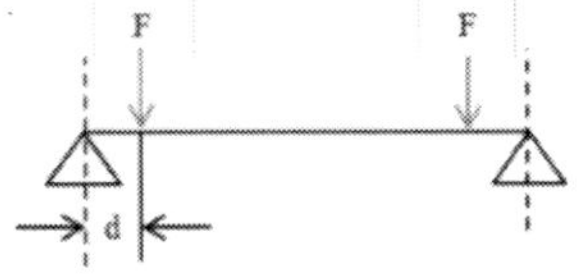

Figure 3. Schematics of the simple FE model. F represents the applied forces, and d is the distance between the inner and outer rollers.

Based on the energy range to be covered by the Si (111) crystal (4.5 - 8.3 keV), the distance from the source to the bent crystal (36 m), and the distance to its focus (18 m), one finds that the sagittal radius needs to be changed between 6.8 m (at 8.3 keV) and 12.6 m (at 4.5 keV).

The divergence of the x-ray beam is 50 μrad (h) x 400 μrad (v), which gives a beam footprint size at the crystal of up to 7.6 mm horizontal and 14.4 mm vertical. Based on the size of the beam footprint and the optimal aspect ratio of 2.36 for simply supported plates [5], an initial set of crystal dimensions was selected as 236 mm long, 100 mm wide between the supporting edges, with a thickness of 1 mm.

With the above-mentioned crystal parameters, a sagittal radius of 6.8 m and the equation for the analytical solution given in Ref. [5], we obtained 0.2205 N-m for each of the two moments. These moments were implemented in the simplified FE model as two lines of 22.05 N force each simulating the rollers located at 10 mm from the pivot points. The lines of force had to be increased to 22.31 N to obtain the correct radius of curvature in the FE model.

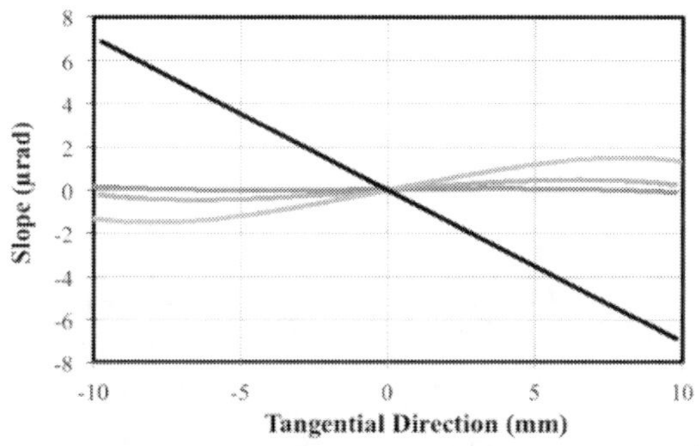

Figure 4. Tangential slope error along the middle of the sagittally bent crystal. The red line is for a crystal length of 236 mm bent to a sagittal radius of 6.8 m by applying idealized lines of force (no rollers) while the black line shows the substantial slope error in the same if rollers are used. The slope error can be reduced (blue line) by adjusting the length of the crystal to an optimal value of 274 mm. This optimal solution is valid for a broad range of sagittal radii exemplified by the green line that shows the tangential slope when the crystal is bent to a radius of 12.6 m.

The main result of the simplified FE model is illustrated in Fig. 4. The red trace in the figure shows that the tangential slope error along the middle of the crystal is very small over the illuminated crystal length. However, as shown in the next section, the realistic crystal/roller contact interface has a significant effect on the anticlastic deformation.

3. Roller/crystal interface in the FE model

To properly model the bent crystal, the crystal/roller interface must be addressed. Four cylindrical bodies were added in the FE model: two pairs of stainless steel rollers with a 10-mm diameter. The initial crystal dimensions were taken from the results of the previous FE model that used idealized boundary conditions. In the complete model (illustrated in Fig. 5), the length of the crystal is kept at 236 mm, the distance between the outer rollers is 100 mm, and they are located 5 mm from the ends of the 110-mm-wide crystal. To bend the crystal to the desired radius, forces are applied to the inner rollers, which are located 10 mm from the outer ones. The contact between the rollers and the crystal plate is assumed to be frictionless.

Figure 6 shows that with the optimized forces (22.31 N) obtained in section 2, the anticlastic bending given by the realistic model separates the rollers from the crystal over a large fraction of its length. This has a significant impact on the tangential slope error as demonstrated by the black trace in Fig. 4. This slope error translates to a tangential convex radius of curvature equal to 1.4 km, creating a virtual image of the horizontal source 6 m downstream of its original position.

Despite the separation between the rollers and the crystal, a significant reduction in the tangential slope error can be obtained by optimizing the crystal length while the other parameters in the model is kept unchanged. This is shown by the blue trace in Fig. 4, which was obtained by increasing the crystal length to 247 mm and by applying forces equal to 28.3 N to obtain a 6.8-m sagittal radius.

As mentioned in section 2 the monochromator energy range requires varying the sagittal radius between 6.8 and 12.6 m. The latter radius can be obtained by simply reducing the couple of forces to 15.2 N while all the other parameters in the FE model are unchanged. As expected, the slope errors along the tangential direction for 12.6 m, shown by the green trace in Fig. 4, are smaller than for the 6.8-m sagittal radius.

4. Ray tracings

The ray tracings were performed with the SHADOW code [6] with the timing slit set to 25 μm. The chirped source, a monochromator, and a KB mirror pair were included in the ray tracings. The KB set is assumed to have ideal figures so that the ray tracing reveals the result of aberrations in the bent crystal. Figure 7 shows the ray tracings at the sample position assuming the crystal is a perfect sagittal cylinder with a 6.8-m radius. Figure 8 shows the ray tracings including the deformation obtained from the FE model described in section 3. The captions

Figure 5. Schematic representation of the FE model with contact. The outer rollers are fixed, and a force is applied to the inner rollers to achieve the desired bend radius.

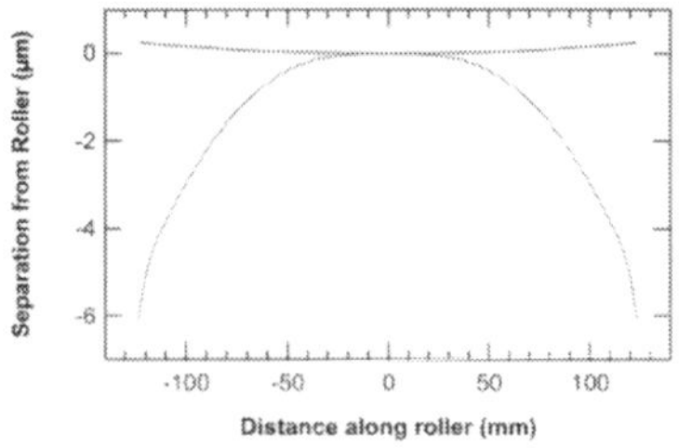

Figure 6. Crystal deformation along a pair of rollers. Red: inner roller; Blue: outer roller.

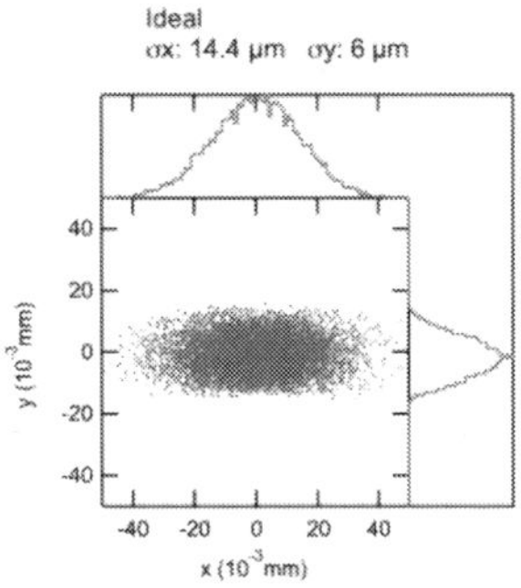

Figure 7. Ray tracings at the sample without deformation.

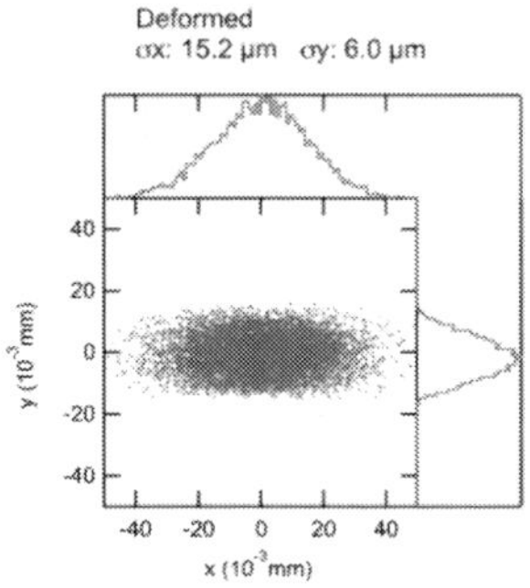

Figure 8. Ray tracings at sample with anticlastic deformation.

above the ray tracings show that the horizontal standard deviation increases only slightly, from 14.4 μm to 15.2 μm.

5. Conclusions

The proper modeling of the roller/crystal interface leads to non-uniform contact pressure and allows separation between the rollers and plate. Determining an optimal aspect ratio (other than the one predicted by the idealized model) can significantly reduce the anticlastic deformation. The optimized crystal geometry of the bendable crystal of the SPXIM beamline has dimensions of 247 mm long x 110 mm wide with a thickness of 1 mm. Ray tracings show that the optimized geometry and sagittal bending based on four rollers has a minimal effect on the focal spot.

Acknowledgments

The Advanced Photon Source, an Office of Science User Facility operated for the U.S. Department of Energy (DOE) Office of Science by Argonne National Laboratory, was supported by the U.S. DOE under Contract No. DE-AC02-06CH11357.

References

[1] Reininger R, Dufresne E M, Borland M, Beno M A, Young L, Kim K-J, and Evans P G 2013 *Rev. Sci. Instrum.* **84** 053103.
[2] Kirkpatrick P and Baez A V 1948 *J. Opt. Soc. Am.* **38** 766.
[3] Sparks C J Jr., Ice G E, Wong J, and Batterman B W 1982 *Nucl. Instrum. Methods Phys. Res.* **194** 73.
[4] Batterman B W, Berman L 1983 *Nucl. Instrum. Methods Phys. Res* **208** 327.
[5] Kushnir V I, Quintana J P, and Georgopoulos P 1993 *Nucl. Instrum. Methods Phys. Res. A* **328** 588.
[6] Welnak C, Chen G J, and Cerrina F 1994 *Nucl. Instrum. Methods Phys. Res. A* **347** 344.

Application of CHESS single-bounce capillaries at synchrotron beamlines

Rong Huang[1], Thomas Szebenyi[1], Mark Pfeifer[1], Arthur Woll[1], Detlef-M. Smilgies[1], Kenneth Finkelstein[1], Darren Dale[1], Yuxin Wang[2], Joan Vila-Comamala[2], Richard Gillilan[3], Michael Cook[3] and Donald H. Bilderback[1,4]

[1]Cornell High Energy Synchrotron Source, Cornell University, Ithaca, NY14853, USA
[2]Advanced Photon Source, Argonne National laboratory, Lemont, IL 60439, USA
[3]MacCHESS (Macromolecular Diffraction facility at CHESS), Ithaca, NY14853, USA
[4]School of Applied Engineering Physics, Cornell University, Ithaca, NY 14853, USA

Email: rh66@cornell.edu

Abstract. Single-bounce capillaries are achromatic X-ray focusing optics that can provide efficient and high demagnification focusing with large numerical apertures. Capillary fabrication at CHESS can be customized according to specific application requirements. Exemplary applications are reviewed in this paper, as well as recent progress on condensers for high-resolution transmission X-ray microscopy and small focal size capillaries.

1. Introduction

Ellipsoid shaped single-bounce glass capillaries fabricated at CHESS have been used as achromatic X-ray focusing optics for various applications at synchrotron beamlines. Such capillaries provide high demagnification and a single short focal length in both transverse directions. The 360 degree azimuthal acceptance angle provides a large numerical aperture for an achromatic X-ray optic in comparison to mirrors. Capillary fabrication at CHESS can be customized according to specific application requirements. Single-bounce capillaries support a wide range of applications, including microbeam X-ray fluorescence [1] and tomography [2], confocal X-ray microscopy [3.4], microcrystallography [5], microbeam small-angle X-ray scattering (μSAXS) [6,7], microbeam high-resolution X-ray diffraction and standing wave technique [8], high pressure studies with diamond anvil cells [9], microbeam grazing-incidence wide-angle X-ray scattering (μGIWAXS) [10], full-field transmission X-ray microscopy (TXM), X-ray emission spectroscopy, etc. Exemplary applications will be reviewed in following from the focusing optics perspective.

2. Exemplary applications of CHESS single-bounce capillaries

2.1. Using capillaries as primary focusing optics

The quality of capillary optics is well matched for a variety of applications at CHESS. Currently, CHESS typically operates in a dual-beam (electron and positron) mode. As a result, the FWHM source sizes at CHESS are approximately 1 mm vertically, and in a range of 1.5 to 3.5 mm horizontally. CHESS beamlines vary in length over a range of 12 – 42 m. For example, with a source size of 2 mm and source distance of 20 m, a 5 μm focus would require a demagnification of 400 and a 50 mm focal length, which is well matched to the capabilities of capillaries. Given an angular source size of approximately 100 μrad, capillary imperfections will not significantly degrade focusing performance if the slope errors are less than about 50 μrad. Capillaries can be produced with larger numerical apertures (NA) than typical KB mirrors, thus optimized for high-flux applications. Single-bounce capillaries have been used as micro-focusing optics for most microbeam experiments at CHESS in the last decade. A CHESS capillary at DESY has also been used for many applications since it was first provided in 2003 [2,4,9,11].

2.2. *Using capillary as secondary focusing optics*

The three protein crystallographic beamlines at CHESS (A1, F1 and F2) all have horizontal and vertical focusing. The beam sizes at sample positions are in the range of 0.2 – 0.3 mm (V) × 1 – 2 mm (H), with focusing divergences of about 0.5 mrad vertically and 0.5 – 2 mrad horizontally. With such a divergent beam, it is impossible using a simple aperture to make a microbeam smaller than 20 μm. However, a focus smaller than 20 μm has been achieved using capillaries as secondary focusing optics [5,12]. Moreover, the capillary slope errors are negligible in these setups because they are smaller than the angular size of the beam intercepted by the capillary [12].

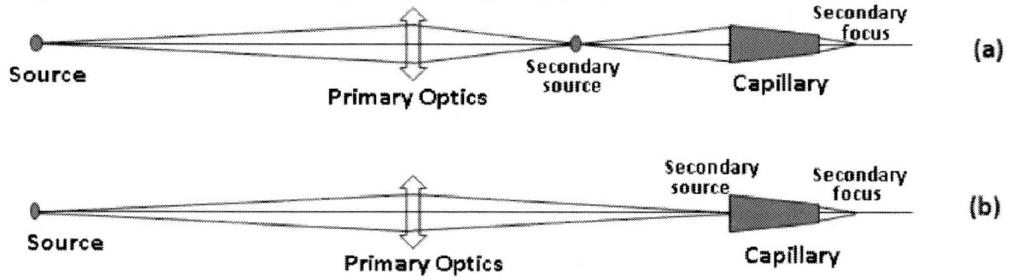

Figure 1. (a) The commonly used secondary focusing configuration with secondary source upstream of secondary optics. (b) With secondary source on secondary optics such as a capillary, a microbeam is focused with maximized flux acceptance.

Single-bounce capillaries have also been used at 3rd generation sources as secondary focusing optics, taking advantage of their large focusing NA to produce high flux microbeams. Generally speaking, capillaries are not ideal primary focusing optics at 3rd generation sources, since the slope errors are large relative to the angular size of the source. However, the tolerances for secondary focusing optics are relaxed due to the increased angular size of the secondary source. Fig. 1(a) shows the commonly used secondary focusing configuration at 3rd generation beamlines, often with an aperture to redefine the secondary source for downstream optics. An alternative configuration may be used, as shown in fig.1(b), with the secondary source at the capillary to maximize flux acceptance. With this configuration, the secondary focal size is $\sigma = F * \theta$, where F is the focal length of secondary optics and θ is the incident beam divergence [13].

A demonstration experiment has been done at the Advanced Photon Source (APS), beamline 18ID (BioCAT), where the radiation from APS undulator A was focused horizontally by a sagittal crystal and vertically by a bimorph mirror. With a CHESS capillary located at the secondary source, 42% of all incident flux was successfully focused into a 10 μm spot [14]. This specific capillary, however, had significant figure errors, yielding noticeable structure in the far-field image. A capillary with negligible slope errors relative to the 90 μrad (H) by 60 μrad (V) angular size of the secondary source would be expected to produce a 3 μm focal spot [13] at this setup. The focused flux could also be increased if a) the beamline primary optics were improved to reduce primary focus size, or b) a capillary with larger inside diameter commensurate with the primary focus were fabricated.

2.3. *X-ray condenser for transmission X-ray microscopy*

A capillary is an excellent optical element to provide incoherent illumination in a TXM system using a Fresnel zone plate (FZP) as objective lens. The resolution of hard X-ray TXMs at synchrotron beamlines has improved with the availability of high-resolution FZPs (which is constrained by the outermost zone width). APS 32ID uses a high-resolution objective FZP based on the zone doubling technique [15], with an outermost zone width of 20 nm [16]. Compared to a previous 40 nm FZP, the NA of the condenser must be doubled, increasing the full divergence from 3.5 mrad to 7 mrad. A CHESS 8 mrad full-divergence capillary was provided to APS recently to match the 20 nm FZP, and the 30 nm structure of a test pattern was easily resolved with this TXM setup (fig.2). Using an integrated circuit test sample and the same beamline conditions and exposure times, a comparison experiment was also done between the 2 TXM configurations: a) 40 nm zone width FZP with a 3.5 mrad condenser and b) the 20 nm zone width FZP with a CHESS 8 mrad capillary condenser. The image taken with the 20 nm zone width FZP and 8 mrad capillary has better illumination uniformity and contrast than the image taken with 40 nm FZP and 3.5 mrad condenser.

Figure 2. The 30 nm fine structure of a test pattern can be easily resolved with the TXM setup with APS 20 nm outmost zone width FZP together with CHESS 8 mrad capillary condenser.

2.4. Miniature toroidal mirrors

The CHESS C-line source-to-sample distance is 14.5 m and beam size at the sample is typically set to 1 mm (V) × 2 mm (H) using 1-to-1 vertical focusing and horizontal slits. A special 250 mm long capillary, with entrance diameter of 2.21 mm, was designed to increase the flux density at the sample as well as enhanced the energy resolution for analyzer-based X-ray emission spectroscopy applications by reducing the emitting source dimensions. The optic was aligned to illuminate the bottom 30% of the capillary inner surface, thus about 35% of the 1 mm by 2 mm incident beam cross section was accepted and focused to a 50 µm spot.

Capillaries are also used as miniature toroidal mirrors at CHESS to create microbeams for scattering applications. For µSAXS, the focused beam divergence can be controlled using a set of slits upstream of the capillary, in order to limit X-ray illumination to a fraction of capillary inner surface [7]. The focused beam divergence can be measured from the far-field intensity distribution, and a symmetric 1 mrad divergence in both vertical and horizontal directions can be achieved. [7]. For µGIWAXS, the divergence control is less strict, and often it suffices to limit capillary acceptance to about a quarter of its full illumination. In both cases a typical beam size was 15 µm using an 8 mrad capillary with a gain of 700 and a working distance of 55 mm. For higher resolution, down to 10 µm, a 4 mrad capillary with 35 mm working distance was used [10].

3. Current capillary development work at CHESS

CHESS is planning upgrades that will yield significant improvements to the source emittance. The quality of capillary optics needs to improve to keep pace with the expected improvements to the CHESS source. Years ago at ESRF, a sub-micron focal size was achieved using a capillary with a short 0.6 mm working distance [17]. At CHESS we are working to improve our capillary fabrication capability, with a goal of 1 µm focal size and multi-mm working distance. The smallest focal sizes we have achieved to date are 2.8 µm using a partially-illuminated capillary with 12 mm working distance [18], and 4 µm using a fully-illuminated capillary with 17 mm working distance, as shown in fig.3.

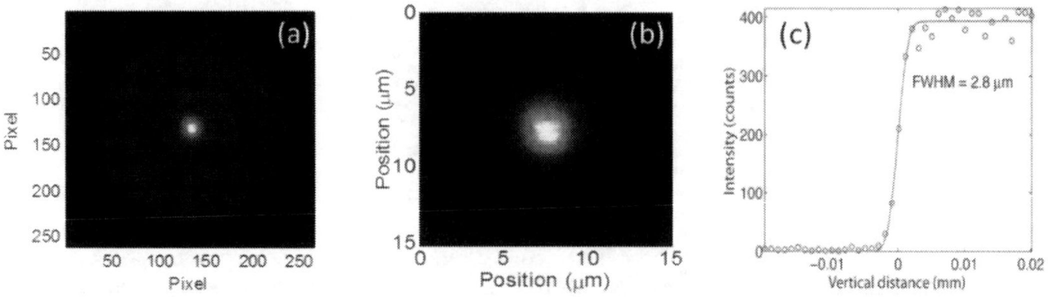

Figure 3. (a) Focused beam and the unblocked direct beam background through the capillary tip opening. The relative intensity of the 75 μm diameter direct beam was magnified by about a factor of 10 at image processing to make it visible in this image without saturating the focused beam. (b) A close look of the 4 μm focal spot by a capillary (labeled AWc_mr5f15_07) with working distance of 17mm. (c) 2.8 μm focal size was measured with partial illumination of a capillary of 12 mm working distance (capillary labeled dsb_mr8f10_07).

We will improve the capillary puller so that high-quality capillaries with smaller slope errors can be fabricated for a large range of dimensions, from small ones optimized for 1 μm focal size to large ones that can be used as miniature toroidal mirrors.

4. Summary

CHESS capillaries are used in a wide range of applications, including recently as a condenser for a high-resolution TXM at the APS. Capillaries are suitable for many microbeam applications, and in some cases they provide significant advantages over alternative achromatic focusing optics. The potentials of capillary development in the near future include high-flux secondary microfocusing, smaller focusing approaching 1 μm with multi-mm working distance, miniature toroidal mirrors for higher acceptance and smaller focus, and TXM condensers.

Acknowledgments

CHESS is supported by the NSF & NIH/NIGMS via NSF award DMR-0936384 and NSF DMR-0807731, and the MacCHESS resource is supported by NIGMS award GM-103485.

References

[1] Limburg KE, Lochet A, Driscoll D, Dale DS and Huang R 2007 *X-ray Spectrum* **36** 336
[2] Lind QC, De Nolf W, Janssens K and Salbu B 2013 *J Environ Radioact* **123** 63-70
[3] Woll AR et al 2006 *Appl. Phys. A* **83** 235-238
[4] Wilke M et al 2010 *J Synchrotron Rad* **17** 669
[5] Gillilan RE, Cook MJ, Cornaby SW and Bilderback DH 2010 *J Synchrotron Rad* **17** 227
[6] Li R., Cornaby S, Kamperman M and Smilgiers D-M 2011 *J Synchrotron Rad* **18** 697
[7] Lamb JS et al 2006 *J Appl Cryst* **40** 193
[8] Kazimirov A, Bilderback DH, Huang R, Sirenko A and Ougazzaden A 2004 *J Phys D* **37** L9
[9] Schmidt C, Rickers K, Bilderback DH and Huang R 2007 *Lithos* **95** 87
[10] Smilgies D-M et al 2013 *Phys Status Solidi RRL* **7** 177
[11] De Samber B et al. 2010 *Powder diffr* **25** 169
[12] Huang R and Bilderback DH 2003 *AIP Conf Proc* **705** 712
[13] Huang R, Meron M, Kujala N and Barrea R 2010 *J Synchrotron Rad* **17** 644
[14] Barrea R, Huang R, Cornaby S, Bilderback DH and Irving TC 2009 *J Synchrotron Rad* **16** 76
[15] Vila-Comamala J et al 2010 Nanotechnology **21** 285305
[16] Vila-Comamala J et al 2012 *J Synchrotron Rad* **19** 705
[17] Snigirev A, et al 2007 *J Synchrotron Rad* **14** 326
[18] Woll AR et al 2012 *Proc of SPIE* **8502** 85020K-1

High-throughput Toroidal Grating Beamline for Photoelectron Spectroscopy at CAMD

O Kizilkaya[1], R W Jiles[1], M C Patterson[2], C A Thibodeaux[3], E D Poliakoff[3], P T Sprunger[1,2], R L Kurtz[1,2], E Morikawa[1]

[1]Center for Advanced Microstructures and Devices, Louisiana State University, Baton Rouge, LA 70806, USA
[2]Department of Physics and Astronomy, Louisiana State University, Baton Rouge, LA 70803, USA
[3]Department of Chemistry, Louisiana State University, Baton Rouge, LA 70803, USA

orhan@lsu.edu

Abstract. A 5 meter toroidal grating (5m-TGM) beamline has been commissioned to deliver 28 mrad of bending magnet radiation to an ultrahigh vacuum endstation chamber to facilitate angle resolved photoelectron spectroscopy. The 5m-TGM beamline is equipped with Au-coated gratings with 300, 600 and 1200 lines/mm providing monochromatized synchrotron radiation in the energy ranges 25-70 eV, 50–120 eV and 100–240 eV, respectively. The beamline delivers excellent flux (~10^{14}-10^{17} photons/sec/100mA) and a combined energy resolution of 189 meV for the beamline (at 1.0 mm slit opening) and HA-50 hemispherical analyzer was obtained at the Fermi level of polycrystalline gold crystal. Our preliminary photoelectron spectroscopy results of phenol adsorption on TiO_2 (110) surface reveals the metal ion (Ti) oxidation.

1. Introduction

The 6 meter toroidal-grating monochromator (6m-TGM) beamline at the Center for Advanced Microstructures and Devices (CAMD) at Louisiana State University was decommissioned in 2008. The 5m-TGM beamline, formally operated at the Electron Stretcher Accelerator (ELSA) Synchrotron Facility, Bonn University, Germany, was relocated to CAMD in 2010 to be installed on the 1.5 GeV CAMD electron storage ring and the same bending magnet port used for the 6m-TGM at CAMD. The beamline commissioning was completed in 2011 and an ultrahigh vacuum (UHV) endstation equipped with a HA-50 electron analyzer was commissioned and attached to the 5m-TGM beamline to facilitate vacuum ultraviolet angle-resolved photoelectron spectroscopy. In this report we present the 5m-TGM beamline performance and the preliminary ultraviolet photoelectron spectroscopy results of phenol adsorption on TiO_2 (110) acquired from this beamline.

2. 5m-TGM Beamline

The 5m-TGM beamline was utilized for x-ray absorption experiments at the Bonn electron stretcher and accelerator ELSA, operated at 2.3 GeV in the storage-ring mode with an average current of 30 mA [1]. Side and top views of the 5m-TGM beamline configuration at CAMD are shown in figure 1. The first focusing mirror is attached to a vacuum chamber outside of the shield wall. The focusing mirror, which has a pitch adjustment, is the first optical element on the beamline and is a water-cooled

Journal of Physics: Conference Series **493** (2014) 012024 doi:10.1088/1742-6596/493/1/012024

ellipsoid. Reflected synchrotron light from the focusing mirror with an incidence angle of 84.5° is focused on the entrance slit. The synchrotron radiation is monochromatized using the toroidal diffraction gratings. The monochromator chamber is equipped with ion-etched, Au-coated gratings with 300, 600, and 1200 lines/mm providing highly resolved monochromatized synchrotron radiation in the energy ranges of 25-70, 50–120 eV, 100–240 eV, respectively. All three toroidal gratings are placed on a cradle, which can be moved horizontally with an extended arm operated with a motor outside of the grating chamber, allowing easy grating exchange. The 5m-TGM beamline does not have a refocusing mirror after the exit slit, and this contributes to the high-throughput nature of the beamline. The monochromatic light from the grating is parallel to the floor and produces a beam-spot size of 5 mm x 2 mm at the sample position with a 1mm vertical and 10 mm horizontal openings on the entrance and exit slits.

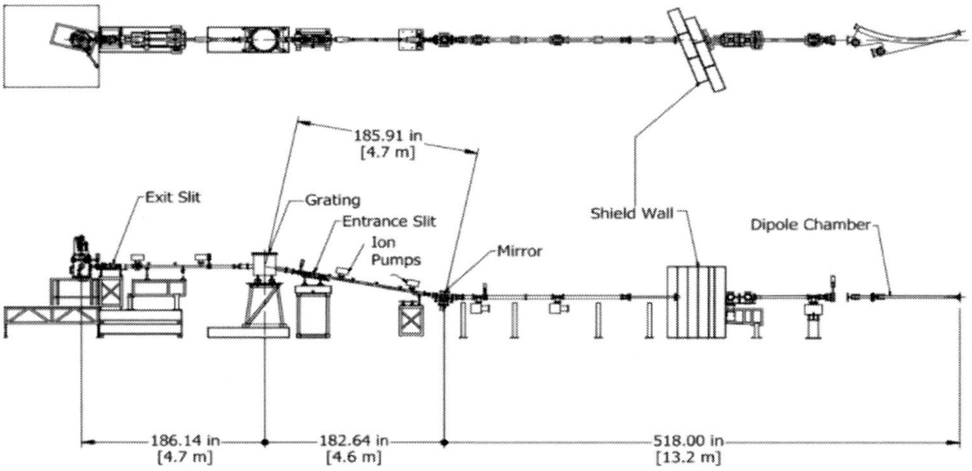

Figure 1. Layout of 5m-TGM beamline at CAMD

The attached ultrahigh vacuum (2×10^{-10} Torr) endstation is primarily used to conduct angle-resolved photoelectron spectroscopy (ARPES) experiments on a variety of materials. The chamber is equipped with a sample load-lock, ion sputtering, and numerous evaporation sources. The sample manipulator allows for full rotation (2-axis) and translation (x-y-z), and facilitates sample temperatures between 90-1600 K. In the case of ARPES data acquisition, photoelectrons emitted from a sample in the endstation are collected with a VSW HA-50 hemispherical electron analyzer, which rotates along 2-axes. This allows for full Brillouin mapping with differing polarizations. The voltages applied to the electrostatic lenses of the analyzer are controlled by a VSW-HAC 300 power supply. The VSW-HAC 300 controller floats over a Valhalla 2701C programmable DC voltage supply controlled by a GPIB interface to set the kinetic energy range of a scan. The charge-signal pulse from a channeltron is capacitively coupled into a voltage signal pulse with an electronic circuit placed outside the vacuum chamber. The pulses are amplified and collected by a timer/counter. A normalization current, proportional to the incoming flux, is also available. A data acquisition program written in LabVIEW program at CAMD records the counts and plots the photoelectron spectrum. The LabVIEW program also controls the grating rotation and its move to an intended photon energy position.

Photon flux delivered by each grating was measured with a GaAsP diode placed at the sample position. Absolute quantum efficiency of the diode was used in the photon flux calculations shown in figure 2a. The low energy grating delivers a high photon flux ~ 10^{17} photons/sec/100mA and high energy grating provides a moderate flux of 10^{14}-10^{15} photons/sec/100mA. In order to determine the combined beamline and analyzer resolution, a Fermi edge of a clean polycrystalline gold crystal was measured at 40 eV photon energy. The full width half maximum of the Fermi edge, 215 meV (see figure 2b), demonstrates a moderate resolution. By correcting for temperature broadening (~4kT=103 meV at 300K), the combined resolution of 189 meV is obtained with a 1mm entrance and exit slit setting of the beamline. It should be added that substantially better combined resolution will be achieved with decreased slit widths.

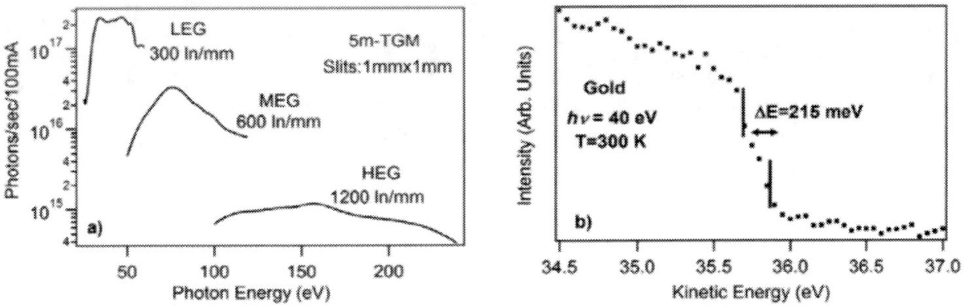

Figure 2. a) Photon flux of LEG, MEG and HEG at the sample position using a GaAsP diode. b) Full width half maximum of the Fermi edge of polycrystalline gold.

3. Phenol adsorption on TiO_2(110)

Recent studies demonstrated that airborne fine and ultrafine particles are mostly generated in combustion sources and these particles cause cardiopulmonary disease and even may trigger cancer [2]. These studies also indicated that airborne particles also contain environmentally persistent free radicals (EPFRs) generated in combustion reactions. The Dellinger group reported that their electron paramagnetic resonance (EPR) experiment demonstrated the formation of environmentally persistent free radicals on metal oxide nanoparticles [3]. Precursors such as phenol initially physisorb on Cu(II)O nanoparticles supported on silica and chemisorption takes place via H_2O elimination through an electron transfer mechanism from phenol to metal oxide resulting in the reduction of metal ion (Cu(I)O).

Electron energy loss spectroscopy (EELS) and EPR spectroscopy experiments were conducted by our group to unravel the mechanisms of EPFR formation from phenol precursors on the surface of well-characterized metal oxides TiO_2 (110) and ultrathin films of Al_2O_3 grown on NiAl (110) [4]. Neither of the techniques decisively determined the metal oxide reduction after radical formation. On the other hand, EELS data from phenol adsorbed on TiO_2 indicated that phenol adsorbs the commonly known defect sites. This observation is proposed as a possible mechanism for the oxidation of the metal oxide substrate.

To elucidate the electronic structure and the interaction of phenol with TiO_2 (110), we performed X-ray and vacuum ultraviolet (VUV) photoelectron spectroscopy measurements. X-ray data was collected in an UHV chamber equipped with an Omicron XM1000 X-ray source and SPECS PHOIBOS 150 hemispherical electron analyzer in the Department of Physics at LSU. A clean TiO_2 (110) surface with a small amount of oxygen vacancies (reduced surface) was prepared by a repeated cycles of Ne ion sputtering (Ne^+, 1.5 kV, 10μA) and annealing (10 minutes at 650 ^{0}C) in the UHV vacuum. The presence of oxygen vacancies/defects on the clean TiO_2 (110) surface leads to an excess of electrons in the surrounding areas of the defect sites. The formal oxidation state of the neighboring titanium cations is reduced from 4+ to 3+. X-ray photoelectron data (not shown here) undertaken at room temperature adsorption of phenol on a reduced TiO_2 (110) surface demonstrated that Ti 3+/Ti 4+ ratio decreases after dosing the surface with phenol. To corroborate this finding, we employed VUV photoelectron spectroscopy experiment at our new 5m-TGM beamline at CAMD. The phenol dosage was facilitated by introducing sublimated vapor into the UHV systems through a leak valve holding a pyrex tube loaded with solid phenol.

Figure 3 shows the photoelectron spectra of both clean and 100 L phenol exposed TiO_2 (110) surface. Both spectra were collected with photon energy of 56 eV, which is within the range of the Ti resonance energies (3p-4s transition) to enhance the intensity of Ti defect states. On the clean photoelectron spectrum (figure 3a), oxygen vacancies induce a defect state that appears at 1 eV binding energy. The features between 3-9 eV are mainly O-2p states hybridized with Ti-3d states. Upon phenol adsorption new electronic features appeared on the valence band at a binding energies of 2.4 eV, 10.7 eV, 13.5 eV and 16.7 eV as seen in the figure 3b. Very similar features were also observed for catechol adsorption on TiO_2 (110) [5]. The structural changes of the O-2p band indicate phenol derived molecular orbitals existence in the binding energy range of 3-9 eV. The defect state appearing at 1 eV binding energy on the clean reduced TiO_2 (110) has almost completely vanished after dosing 100 L phenol at RT. VUV photoelectron spectroscopy results indicate that phenol adsorbs on the surface of TiO_2 and occupies the oxygen vacancies sites, which supports the X-ray photoelectron results we independently performed. It is also suggested that phenol adsorption on TiO_2 (110) presumably oxidizes the metal ion (Ti). We are currently performing theoretical calculations to compare with the results obtained from X-ray and VUV photoelectron spectroscopy measurements.

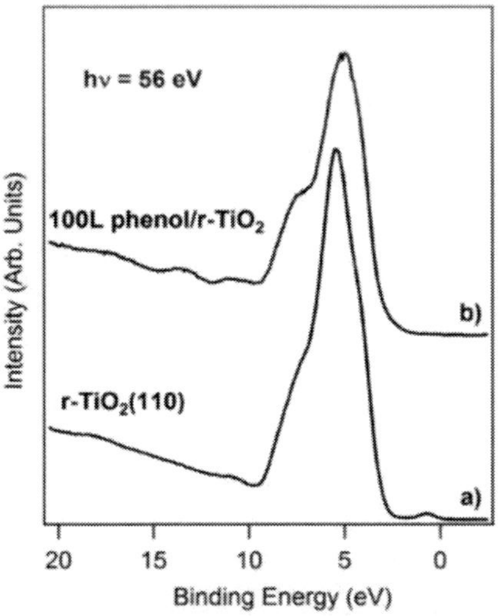

Figure 3. Photoelectron spectra of clean (a) and 100L phenol exposed (b) TiO_2 (110).

Acknowledgements
We acknowledge support from the National Institute of Environmental Health Sciences Superfund Research Program grant number P42 ES013648-03.

References
[1] Pavlychev A A, Franke R, Bender S and Hormes J 1998 *J. Phys.: Condens. Matter* **10** 2181
[2] Pope C A, Burnett R T, Thun M J, Calle E E, Krewski D, Ito K and Thurston G D 2002 *J. Am. Med. Assoc.* 287 1132
[3] Lomnicki S, Truong H, Vejerano E and Dellinger B 2008 *Environ. Sci. Technol.* **42** 4982
[4] Patterson M C, Keilbart N D, Kiruri L W, Thibodeaux C A, Lomnicki S, Kurtz R L, Poliakoff E D, Dellinger B and Sprunger P T 2013 *Chem. Phys.* accepted
[5] Rangan S, Theisen J, Bersch E and Bartynski R A 2010 *Appl. Surf. Sci.* **256** 4829

Simulations of a FIR Oscillator with Large Slippage parameter at Jefferson Lab for FIR/UV pump-probe experiments

S V Benson[2], L T Campbell[1,3], B W J McNeil[1], G R Neil[2], M D Shinn[2], and G P Williams[2]

[1]Dept. of Physics, University of Strathclyde, Glasgow G4 0NG, Scotland
[2]Thomas Jefferson National Accelerator Facility, Newport News, VA 23606 USA
[3]AsTeC, STFC Daresbury Laboratory and Cockcroft Institute, Warrington WA4 4AD, UK

felman@jlab.org

Abstract. We previously proposed a dual FEL configuration on the UV Demo FEL at Jefferson Lab that would allow simultaneous lasing at FIR and UV wavelengths [1]. The FIR source would be an FEL oscillator with a short wiggler providing diffraction-limited pulses with pulse energy exceeding 50 microJoules, using the exhaust beam from a UVFEL as the input electron beam. Since the UV FEL requires very short pulses, the input to the FIR FEL is extremely short compared to a slippage length and the usual Slowly Varying Envelope Approximation (SVEA) does not apply. We use a non-SVEA code [2] to simulate this system both with a small energy spread (UV laser off) and with large energy spread (UV laser on).

1. Introduction

Though subpicosecond lasers are routinely used in research, many experiments require the use of two synchronized lasers with very different wavelengths. We previously proposed using a far infrared (FIR) laser synchronized to an ultraviolet/vacuum ultraviolet (UV/VUV) oscillator in order to allow experiments to be carried out in which a sample is pumped in the UV or VUV wavelength range and then probed by FIR wave radiation after a known delay. In this way the photo-emitted electron wavefunction can be encoded and studied as was done, for example, in experiments on Kr at the FLASH FEL in Germany [3]. However, the longer wavelength is sufficiently powerful to be the pump as well, and thus one could envisage experiments similar to the pump-probe photoemission study of strongly correlated charge-density-wave materials [4,5], but with the additional flexibility of the wider wavelength range as well as the tuning offered by an FEL. In our earlier study, it was found that the standard FEL oscillator codes used to model the FIR FEL were not appropriate for this latter case due to the extreme violation of the Slowly Varying Envelope Approximation (SVEA). In addition, only one-dimensional simulations were done for the UVFEL so the input parameters for the FIR FEL were not well known. This work shows some of the simulations of both lasers with more appropriate codes and shows that the performance is still reasonable. In fact, the SVEA codes give surprisingly good predictions of the gain. The power output from the non-SVEA codes is smaller but still quite usable.

2. Description of the FEL facility

The UV Demo FEL facility is described in some detail in reference [6]. UV operation reported in that publication was for operation at low charge (60 pC) in order to provide a geometric emittance less than the laser wavelength divided by 4π. For operation with a FIR laser after the UVFEL it helps to have higher charge. Simulations done for this publication assumed 135 pC, which is the standard value for the IRFEL at Jefferson Lab.

Free-electron lasers naturally emit on-axis coherent radiation at the odd harmonics of the resonant wavelength. When the FEL is operated at 3 eV, for example, there is light emitted at 9 eV and 15 eV. At these photon energies there are very few transparent materials, so the best way to take advantage of these VUV harmonics is to lase with an output coupler in which a hole is drilled. The harmonic radiation is then transmitted through the hole and to a user station using reflective optics. The output coupling can be varied in this setup by changing the radius of curvature of the upstream mirror. Emission in the VUV has already been demonstrated on the UV Demo laser at 10 eV [7].

2.1. Far Infrared Laser Layout

Because of a specific user request we did our early studies of this system with the UV laser at 3.3 eV (372 nm wavelength) and the FIR laser at 40 meV (31 microns). To allow the introduction of a FIR FEL downstream of the UVFEL we intend to modify the downstream chicane magnets so that the electron beam travels along a zigzag path and the dispersion in the middle of the insertion is small (<3 cm). The basic layout is sketched in figure 1.

The angle of the wiggler with respect to the UV beam path is only 25 mrad. This offsets the mirrors of an 8 meter resonator by 10 cm, which is just sufficient for clearance of a 10 cm diameter resonator mirror. The vacuum chamber for the wiggler is approximately 3 meters long so the separation between the UV mode and the electron beam is 38 mm at either end of the vacuum chamber. This is then the maximum dispersion in the FIR FEL.

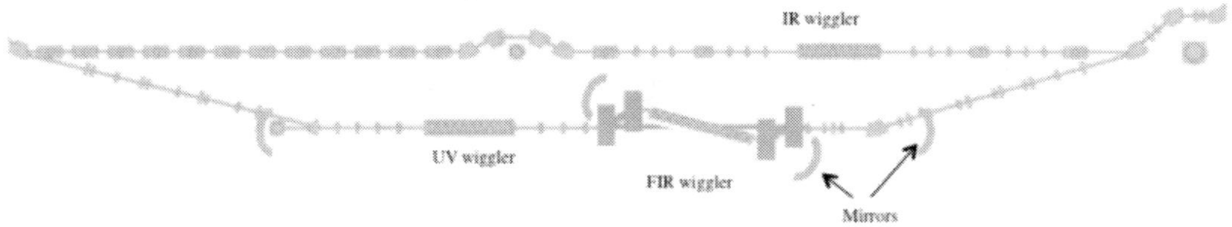

Figure 1. Proposed layout for the FIR FEL downstream of the UV wiggler. Note that the angles in the zigzag are greatly exaggerated and that the actual bend angle would be vertical.

3. Simulations

The parameters for the electron beam, optical resonator and wiggler for the FIR-FEL are given in table 1. The laser output was assumed to be out-coupled through a hole in the downstream mirror. The gain increases with decreasing wavelength so hole-coupling is better matched to the wavelength variation than edge coupling since reducing the wavelength increases both the gain and the output coupling. The resonator at 31 microns is an open resonator with very small diffractive losses. At longer wavelengths, one might have to go to a waveguide resonator to better confine the mode.

The FIR laser was simulated using a one-dimensional non-SVEA FEL code called Puffin [2] for two different energy spreads. The first was for when the UV laser is not lasing. This is useful for getting the FIR FEL to lase and optimizing the lasing. Once the FIR FEL is optimized, the UV FEL can be turned on and the rms energy spread should increase to ~0.9%.

The UVFEL was modeled using the 4 dimensional codes Medusa/OPC [8] as well as 3D Genesis/OPC [9]. When we operate near the peak of the gain curve using the parameters in Table 1 we find reasonably high gain (>60%) and extraction efficiency (0.5%) with the 135 pC bunch when corrected for slippage effects. The *rms* exhaust energy spread is approximately 0.9±0.1% in agreement with Table 1.

Table 1: FIR-FEL wiggler, resonator and simulation electron beam parameters.

Parameter	Value
Electron energy	135 MeV
Bunch charge	135 pC
Bunch length (*rms*)	0.18 psec
Peak current	300 A
Repetition rate	18.7125 MHz
Energy spread (UV FEL not lasing)	0.5%
Energy spread (UV FEL lasing)	0.9%
Emittance	8 mm-mrad
Wiggler wavelength	20 cm
Number of periods	12
Peak wiggler field	3.5 kG
Rayleigh range	108 cm
Losses	3%
Output coupling	1%
Laser wavelength	31 μm

3.1. Assumptions of the simulation

For the FIR FEL the wavelength is only one sixth of the *rms* electron bunch length. Thus, the SVEA simulations are not appropriate. The simulation also assumes that the pulse is Gaussian in time and this is almost never true for a linac-based accelerator. Initial simulations use a Gaussian distribution for the electrons with the *rms* parameters shown in Table 1, but we have also done UV simulations using the results of Start-to-End (S2E) simulations as well. We can then use the output from this program to find the conditions at the entrance to the FIR FEL. S2E simulations of the beam at 135 pC do show a lot of fine structure on the beam as shown in Figure 2.

This fine structure must be taken into account in any simulation. For the UVFEL, we simulated this case using 83 pC of the 135 pC bunch in a short parabolic current profile with 78 fsec full width, 0.6% energy spread, and 1600 A peak current. We used a 4D simulation code for this due to the very short bunch length. The exhaust distribution when just this spike is used in the FEL is shown in figure 3. The *rms* energy spread is still 0.9% but there will be much more coherent undulator radiation from this bunch. We have not yet run Puffin using the distribution in figure 2.

For the FIR FEL simulations using Puffin, we assumed simple Gaussian distributions. The gain and power are weakly dependent on the transverse emittance and the energy spread. We do expect the gain and efficiency to be higher when the S2E distribution is used instead of the Gaussian bunch.

Most users are interested in narrow spectral bandwidth and will want to operate on the shorter end of the cavity-length-detuning curve where the spectral brightness is higher. In figure 3 we show the energy distribution at the output of the UVFEL for operation near the peak of the gain curve. The spectral brightness is highest here and the operation is more stable with a hole out-coupler. The gain here is very high and the extraction efficiency is 0.28%. Approximately half of the extracted power exits the hole.

For the FIR FEL we looked at two different cases, one with the UVFEL on and one with it off. We did the simulations near the peak of the gain curve predicted from the 1D theory. This should produce the best spectral brightness. One critical aspect is the small signal gain. If this is much less than the 1D SVEA simulations, then the laser will not start up. In figure 4 we show the gain vs. pass number for the two cases. It can be seen that the gain, though about 15% lower than that predicted by the SVEA codes, is still sufficient to turn on assuming losses can be kept lower than about 3%. Note that gain much larger than losses is not good in the laser because it produces too much energy spread, leading to losses in the downstream transport.

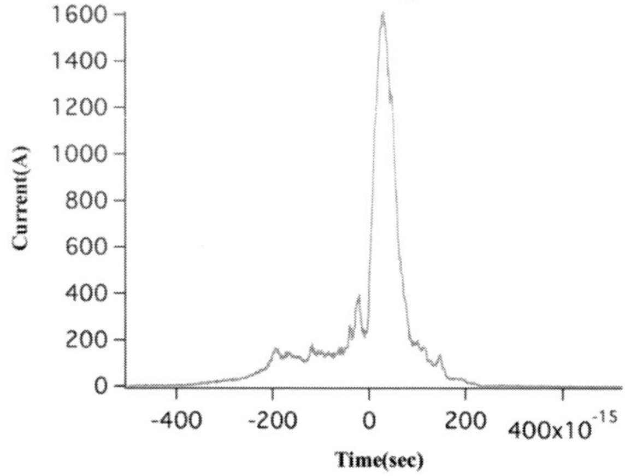

Figure 2. Current profile of a micropulse from an S2E simulation. The rms bunch length is 100 fsec but the FWHM is only 50 fsec.

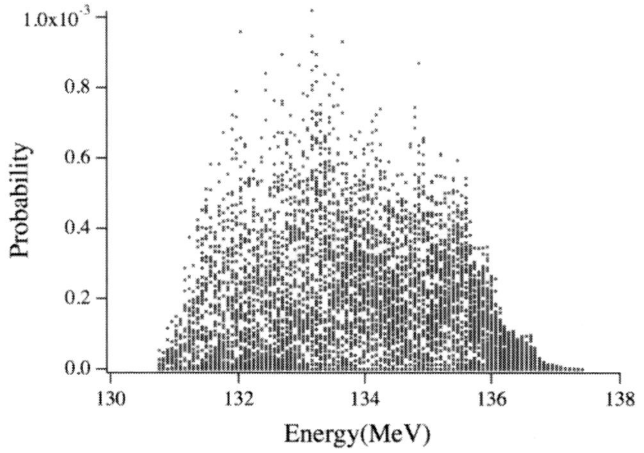

Figure 3. UVFEL exhaust energy distribution as calculated by 4D MEDUSA for the spike present in an S2E simulation (figure 2). The rms energy spread is 0.9% and the energy loss is 0.28%.

In figure 5 we show the saturated distribution with the UVFEL off. The saturated laser pulse is only about 10 optical periods FWHM, which is less than the number of wiggler periods. The spectral bandwidth, however, is only about 3% FWHM, which is very good for such a short bunch. The peak power in the microbunch is a respectable 40 MW. This corresponds to about 750 W of power at the repetition rate of 18.7 MHz. The exhaust energy spread is over 9% full width, which is close to the energy acceptance of the energy recovery arc.

In figure 6 the distributions with the UVFEL on are shown. The peak power drops by about 1/3 but the pulse length and spectral bandwidth are about the same as with the UVFEL on. The micropulse energy is approximately 25 μJ, which is more than enough to pump condensed matter systems.

The predicted gain from Puffin agrees reasonably well with the SVEA codes. The SVEA gain is about 15% more than that predicted by Puffin. The power predicted by Puffin is about half that predicted by the SVEA codes. Though not fatal, this result is interesting and will be explored in future work. As noted above, there is a need to calculate the gain and efficiency for the more realistic S2E phase space distribution.

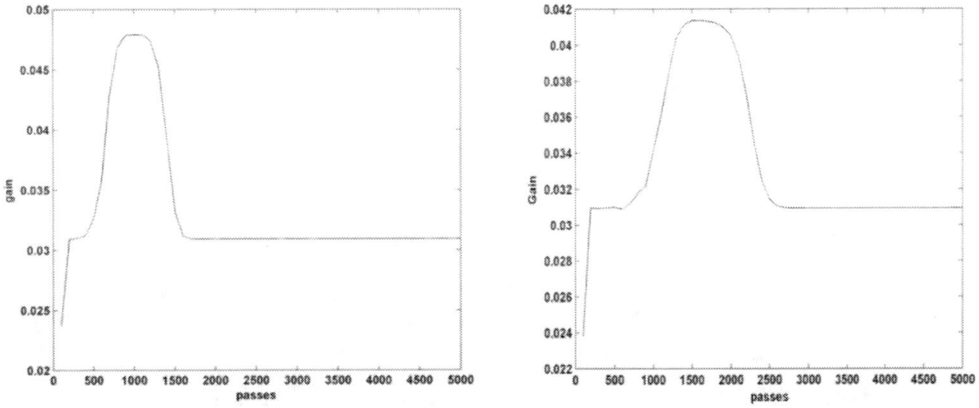

Figure 4. Gain vs. pass number at a cavity length detuning of 2 μm calculated using Puffin. The curve on the left is with the UVFEL off with 0.5% input energy spread. The one on the right is with the UVFEL on with 0.9% input energy spread.

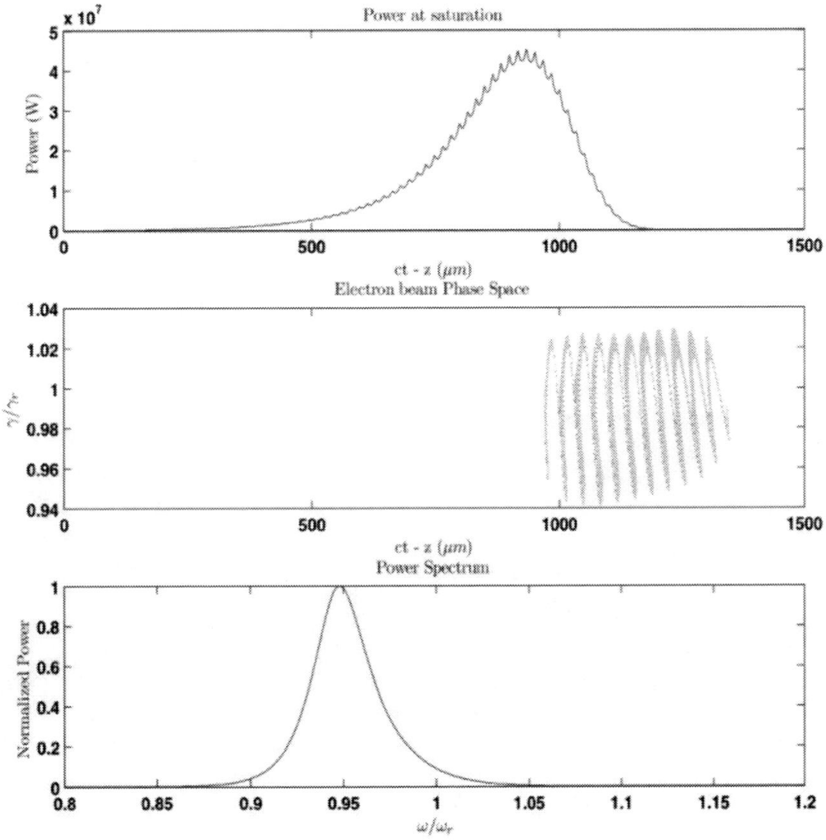

Figure 5. Saturated power (top), electron phase space distribution (middle) and spectrum (bottom) for FIR operation with the UVFEL off.

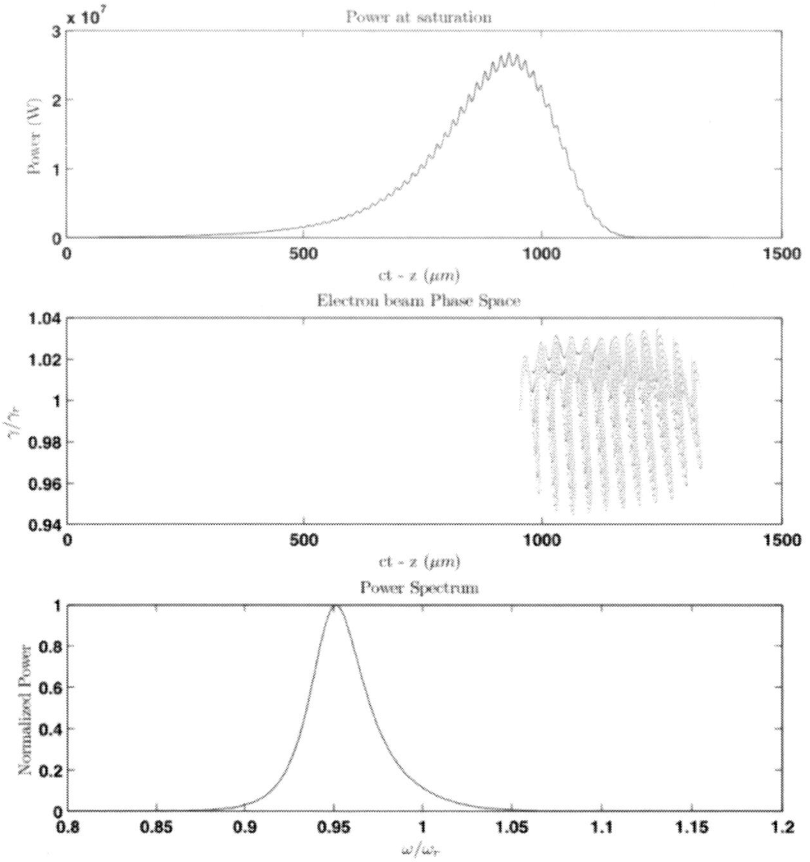

Figure 6. Distributions with the UVFEL on. The simulations so far have assumed simple Gaussian distributions

4. Conclusions
More realistic simulations of a combined VUV/FIR laser source show that the device should work well. Future work will look at the use of S2E simulations to better predict the performance of a physical system. The laser outputs expected for varying electron beam energies are summarized in table 2.

5. Acknowledgment
This work was supported by U.S. DOE Contract No. DE-AC05-84-ER40150, the Air Force Office of Scientific Research, DOE Basic Energy Sciences.

References
[1] Benson S V, Douglas D, Neil G, Shinn M D and Williams G P 2012 *Proceedings of IPAC'12, (New Orleans, LA: JACOW)* p. 1789
[2] Campbell L T and McNeil B W J 2012 *Physics of Plasmas* **19** 093119
[3] Fruhling U et al. 2009 *Nature Photonics* **3** 523

[4] Rohwer T et al. 2011 *Nature* **471**, 490
[5] Smallwood C L Jozwiak C Zhang W and Lanzara A 2012 *Rev. Sci. Instr.* **83** 123904
[6] Douglas D R et al. 2012 *Proceedings of IPAC'12 (New Orleans, LA: JACOW)* p. 2111
[7] Legg R A et al. 2011 *Proceedings of PAC'11 (New York NY: JACOW)* p. 2432
[8] Van der Slot P J M Freund H Miner W H Boller K-J Benson S Shinn M D 2009 *Phys. Rev. Lett.* **102** 244802
[9] Karssenberg J G Van der Slot P J M Volokhine I V Verschuur J W J and Boller K-J 2006 *JAP* **100** 093106

Table 2: Laser output parameters during simultaneous operation.

Parameter	**Value**
UV	
Photon energy	2–5 eV
Bunch length (FWHM)	0.2 psec
Bandwidth	0.3%
Pulse energy	>25 μJ
VUV(third harmonic)	
Photon energy	6–15 eV
Bunch length (FWHM)	0.2 psec
Bandwidth	0.2%
Pulse energy	>10 nJ
FIR	
Photon energy	0.03-0.1 eV
Bunch length	0.4–1.4 psec
Bandwidth	3%
Pulse energy	>25 μJ
All	
Repetition rate	18.7125 MHz

Journal of Physics: Conference Series **493** (2014) 012026 doi:10.1088/1742-6596/493/1/012026

Piezo control for 1 nm spatial resolution synchrotron X-ray microscopy.

K. J. Gofron[1,#], K. Lauer[1], E. Nazaretski[1], H. Yan[1], S. Kalbfleisch[1], A. Greer[2], B. Dalesio[1], Y. S. Chu[1]

[1]National Synchrotron Light Source II, Brookhaven National Laboratory, Upton, NY, USA.

[2]Observatory Sciences Ltd., Cambridge, CB4 0WX, U.K.

[#]E-mail: kgofron@bnl.gov

Abstract. A novel motion control system which utilizes the Power PMAC controller from Delta Tau Data Systems Inc., has been developed for positioning with 1 nm spatial resolution. Present work is a significant step forward towards commissioning of the X-ray microscope which will operate at the Hard X-ray Nanoprobe (HXN) beamline at the NSLS-II. The control system is capable of performing high-speed / high-accuracy on-the-fly scans of the sample with respect to the nano-focusing optics e.g. Multilayer Laue Lenses (MLL) or Fresnel X-ray Zone Plates (ZP) [1]. The Power PMAC controls piezoelectric-based nano-positioning stages using piezo-expansion for short range motion and stick-slip motion for longer travel distances. An EPICS interface to the Power PMAC has been developed allowing for easy integration into a beamline control environment.

1. Introduction

Experiments with nanometer spatial resolution focussed X-ray beams that study phenomena related to elemental composition (via scanning fluorescence microscopy, SFM) or local structure (X-ray diffraction) require accurate positioning of both X-ray nanofocusing optics and samples. During experiments, the sample is raster scanned with respect to the optics. Position stability, down to 1 nm, is required when measurement (e.g. SFM or diffraction) with nano-meter level X-ray beam is performed. Measurements with such unprecedented spatial resolution impose stringent requirements on stability.

In order to achieve the stability requirements, we use active control feedback with optical interferometers to keep track and correct position of nano-focusing X-ray optics (such as MLL and ZP) and sample stage [2,3]. In this paper, we report on the successful use of active feedback loop not only for stationary (non-moving) components of the X-ray microscope, but also active feedback loop for scanning stages. The feedback loop is closed in the Power PMAC controller with quadrature encoder (A quad B) signal generated by Attocube AG interferometers.

2. Experimental arrangement

A typical experiment which can advantageously make use of the actively closed feedback during motion is continuous SFM scan. Unlike a point-by-point scan, during continuous (on-the-fly) scan the scanned/scanning component (sample, or X-ray beam), motion does not stop for acquisition of fluorescence and/or diffraction data. Continuous scans create a substantial speedup of data acquisition.

For instance, a 10 μm x 10 μm fluorescence map (containing 16+ chemical elemental compositions) with pixel resolution of 10 nm (1000 x 1000 pixel image) can be collected within 3 hours with 100 Hz readout, or 20 minutes with 1000 Hz readout. A chemical map, of the same size, acquired by point scan (0.2 s/point) would require 2 consecutive days of synchrotron beamtime (data will be distorted by various instabilities such as injections, beam orbit corrections, etc). Therefore active feedback and shortening data collection time improves stability, and saves user beam time.

The HXN MLL microscope contains about 20 mostly dual (piezo expansion, stick-slip) stages with a total of about 40 motions. Similar type of piezo stages have been used in scanning probe microscopy instruments [4]. The horizontal MLL assembly (used to position the horizontal focusing MLL) has 4 motions (3 translations and 1 rotation). The vertically focussing MML alignment has 5 motions (3 translations and 2 rotations). The Order Sorting Aperture (OSA) implements 3 translation motions and 3 rotations. Finally, the sample stage contains 6 motions. In addition there is a motorized X-ray beam stop, and a motorized tilted mirror is planned for an optical on-axis microscope.

This paper summarizes results of tests performed on the horizontal MLL assembly, where three stacked Attocube AG translation stages were controlled by the Power PMAC and driven by an Attocube ANC-300 piezo driver.

3. Hardware

The control setup consists of a Power PMAC controller, Attocube fiber-optic interferometers, ANC-300 piezo-motor driver, and customized piezo motion stages procured from Attocube AG.

3.1. Motion Controller

The versatility of the Power PMAC makes it an excellent choice for X-ray microscope control. Not only does it have the ability to close a position feedback loop onto a digital A quad B encoder signal, but also on an analog feedback voltage produced by resistive or capacitive sensors (read by a 24-bit ADC UMAC card at servo frequencies). The latter sensors can be used in harmful X-ray environment where traditional optical encoders can be damaged by radiation. Additionally, sin/cos 1 Vpp analog signals are read with 16k multiplication, an increase over 4k multiplication with Delta Tau's previous generation controller, the Turbo PMAC. The piezo expansion is controlled with an accuracy of 18-bit DAC. These high precision features offered by the Power PMAC allow the HXN X-ray microscope to deliver sub-10 nm spatial resolution.

The motion controller consists of a UMAC chassis with a Power PMAC controller card, ACC-24E3 (digital encoder, 18-bit DAC) cards, and ACC-24E2S cards. We added an ACC-36uk (version with 24-bit accuracy) ADC card for voltage driven feedback loops such as capacitive sensors, ACC-14E for TTL control, and ACC-28E for 16-bit ADC. When available, Dsub-15 connector options for the PMAC accessories were purchased for easy connectivity.

3.2. Interferometer

We used two different models of fiber-optic interferometers from Attocube AG. First interferometer is capable of measuring displacement on 12 independent axes powered by a single laser diode (AttoFPsensor). The AttoFPsensor interferometer provides single ended encoder output for each axis and requires conversion to quadrature signal before signal is accepted by motion controller. The newer FPS3010 model requires custom interconnects of encoder quadrature signal from HDMI connector with other DB-15 inputs/outputs of Power PMAC cards.

4. Results

A number of motion tests were carried out to verify the performance of the closed feedback loop. One of the important tests is verification of the absence of parasitic motions. The example of absence of parasitic motion is illustrated in Figure 1, where the Z translation motion does not result in observable translations in X or Y directions (in the stage containing XZY motion axis of MLL horizontal assembly).

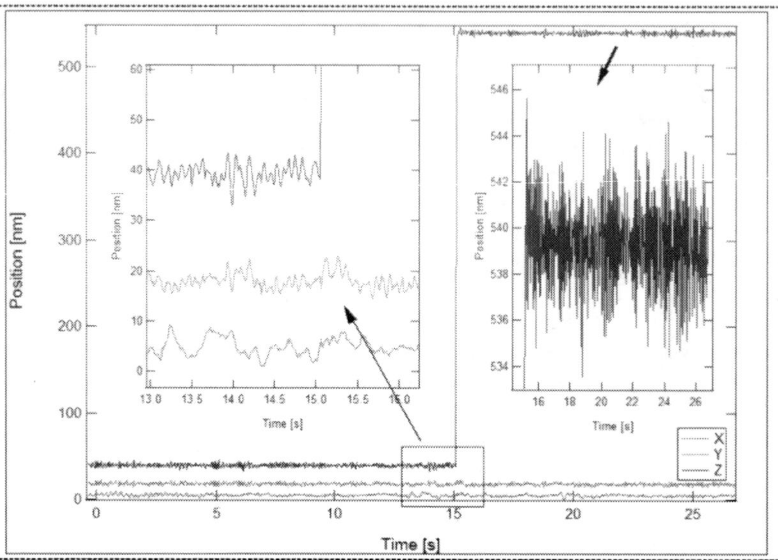

Figure 1. The XYZ stage of the horizontal focusing lens of the MLL microscope shows no XY (green, red) parasitic movement when the Z axis (along X-ray beam) motion is performed (blue plot). The inset plots show details of the position with closed loop interferometer feedback under ambient conditions.

The stacking of the stages in Figure 1, is such that in standard beamline coordinates X (lateral) translation carries the Z (along X-ray beam) translation, which in turn carries the Y (vertical) translation stage.

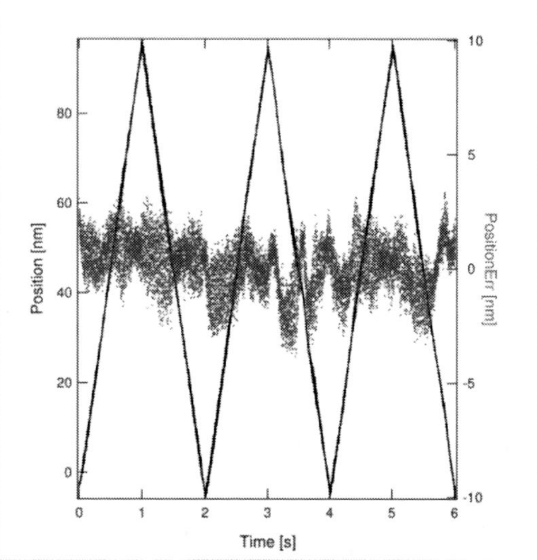

Figure 2. Closed loop piezo-expansion ramp motion of MLL-H X-axis with interferometer feedback implemented. The position error (red/rigth) is within +/- 2 nm of requested position.

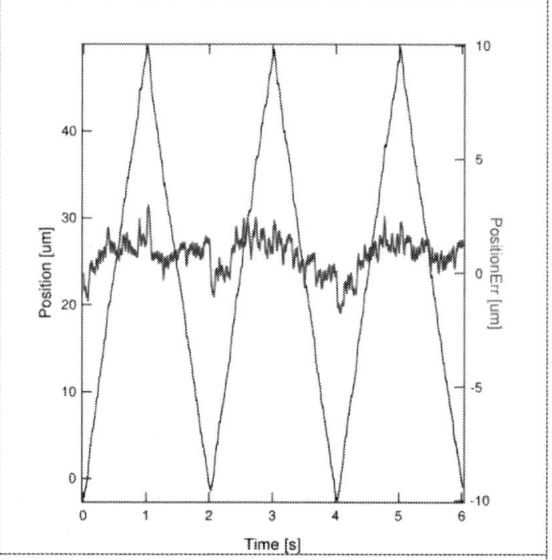

Figure 3. Stick-slip motion is used for long travel range (1.0 µm – 2 cm) of optics components and sample with interferometer feedback. The position error (red/rigth) is within +/- 2 um of requested position.

Figure 2 illustrates a closed feedback loop where ramping is implemented. Such ramping (continuous velocity) motion is typically used for continuous scans, and it is seen that deviation from a requested position does not exceed 2 nm (Fig. 2) during piezo-expansion and about 2 μm (Fig. 3) during stick-slip closed feedback motions, respectively.

5. Control flow

As shown in Figure 4, the quadrature input from an interferometer enters the Power PMAC controller through Digital Encoder Mezzanine board. The stick-slip is controlled by a step/direction output of the encoder board, which for ANC-300 is further converted to CW/CCW pulses (not shown). The piezo expansion is controlled by Acc-24E3 18-bit DAC amplifier output.

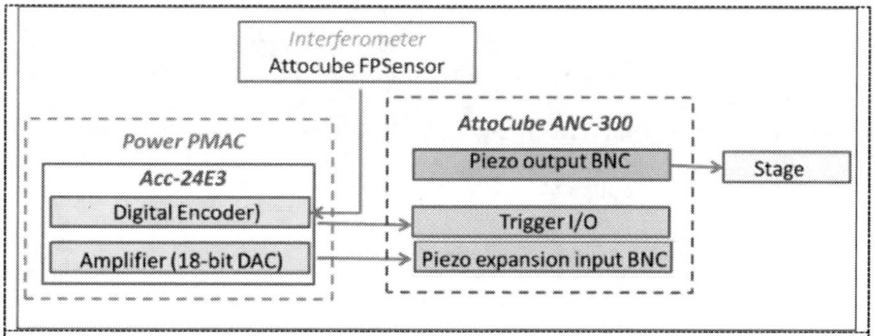

Figure 4. The control flow for the piezo-expansion and stick-slip motions.

6. Conclusions

Recent developments in control technology and piezo motion have made possible continuous scans that improved both stability and substantially reduced data collection time, at the same time providing nm-scale spatial resolution. Integration of different positioning devices (controllers, interferometers, piezo drivers, etc.) required manufacturing of a number of custom transition boards where interferometer signals have been brought together to connectors available on motion controller, and piezo drivers. Nanometer-scale spatial resolution measurements confirmed the validity of the approach used to implement scanning algorithms and corresponding hardware.

7. Acknowledgements

Work at Brookhaven was supported by the Department of Energy, Office of Basic Energy Sciences under contract DE-AC-02-98CH10886.
We thank Dennis Kuhne and James Biancarosa for experimental support. We thank Deming Shu at APS for collaborating on developing the HXN MLL microscope.

8. References

[1] Hanfei Yan and Yong S. Chu, "Optimization of multilayer Laue lenses for a scanning X-ray microscope," Journal of Synchrotron Radiation 20 (1), 89-97 (2013).
[2] Hyon Chol Kang, Hanfei Yan, Yong S. Chu et al., "Oxidation of PtNi nanoparticles studied by a scanning X-ray fluorescence microscope with multi-layer Laue lenses," Nanoscale (2013).
[3] Hanfei Yan, Yong S. Chu, Jorg Maser et al., "Quantitative x-ray phase imaging at the nanoscale by multilayer Laue lenses," Scientific Reports 3, 1307 (2013).
[4] Yongkai Xu, Peter H. Meckl, "Time-Optimal Motion Control of Piezoelectric Actuator: STM Application," Proceeding of the 2004 American Control Conference Boston, Massachusetts June 30 -July 2,2004

Journal of Physics: Conference Series **493** (2014) 012027 doi:10.1088/1742-6596/493/1/012027

Optomechanical design of a high-precision detector robot arm system for x-ray nano-diffraction with x-ray nanoprobe

D Shu[1], S. Kalbfleisch[2], S. Kearney[1,3], J. Anton[1,3], and Y. S. Chu[2]

[1]Advanced Photon Source, Argonne National Laboratory, Argonne, IL 60439, USA
[2]National Synchrotron Light Source, Brookhaven National Laboratory, Upton, NY 11973, USA
[3]University of Illinois at Chicago, Chicago, IL 60607, USA

Email: shu@aps.anl.gov

Abstract. Collaboration between Argonne National Laboratory and Brookhaven National Laboratory has created a design for the high-precision detector robot arm system that will be used in the x-ray nano-diffraction experimental station at the Hard X-ray Nanoprobe (HXN) beamline for the NSLS-II project. The robot arm system is designed for positioning and manipulating an x-ray detector in three-dimensional space for nano-diffraction data acquisition with the HXN x-ray microscope. It consists of the following major component groups: a granite base with air-bearing support, a 2-D horizontal base stage, a vertical axis goniometer, a 2-D vertical plane robot arm, a 3-D fast scanning stages group, and a 2-D x-ray pixel detector. The design specifications and unique optomechanical structure of this novel high-precision detector robot arm system will be presented in this paper.

1. Introduction

A traditional way to perform diffraction measurement is to use a diffractometer. In carrying out diffraction measurement using a small focused beam, or nanodiffraction, using a conventional diffractometer faces three major instrument challenges. First, rotation motion of the sample typically has a large (> 10 microns) run-out error known as a sphere of confusion. Second, the circles for vertical and horizontal diffraction angles are mechanically coupled to the sample, causing unwanted displacement of the sample position when the detector is rotated. Three, some x-ray microscopes used for micro- or nanodiffraction require a large vacuum enclosure, making it virtually impossible to use a conventional diffractometer.

For the x-ray microscope for the Hard X-ray Nanoprobe (HXN) beamline at the National Synchrotron Light Source II (NSLS-II) at Brookhaven National Laboratory, which is designed for x-ray microscopy capabilities with an initial goal for a spatial resolution of 10 nm [1], we take a novel design approach for nanodiffraction. In our design, we completely de-couple the detector positioning system from the sample stage stack and use a high-precision detector robot arm system in generating full spherical coordinate motions for an array detector. These degrees of freedom of motion include the horizontal diffraction angle, the vertical diffraction angle, and the sample-to-detector distance (we shall refer to this as detector distance in the rest of the manuscript). When performing Bragg ptychography [2], adjusting detector distance is important because it defines the maximum range of diffraction angle, which contributes to the spatial resolution limit of the reconstruction.

This paper describes detailed optomechanical design and specifications of the high-precision detector robotic arm system which employs a unique vertical plane robotic arm structure with duo-vertical-stages [3].

2. Design requirement for the detector robot arm system

The instrument requirement for the NSLS-II HXN beamline is to achieve a horizontal diffraction angular range from -5 to 65°, a vertical diffraction angular range from -5 to 45°, and a sample-to-detector distance from 0.25 to 0.54 m (0.25 to 2.5 m with base station positioning). The horizontal diffraction angle is produced by using a 2-D horizontal base stage plus an in-plane goniometer. The vertical diffraction angle is achieved using a vertical plane robot arm structure. In addition, a 3-D fast scanning stages group is used to control the detector distance and the detector raster capability to achieve a larger q-range.

The normal load capacity of the custom-built HXN-Z7-47 detector robot arm is 18 kg, which is not only able to handle the total weight of an array detector, such as a TimepixTM QTPX-262K (Amsterdam Scientific Instruments) or other moderate-size detector, and a pair of motorized stages for the area detector raster scan, but also is capable of carrying out the dynamic forces generated during the detector raster scan.

The initial design goal for the unidirectional positioning repeatability of the HXN-Z7-47 robot arm is 20 microns in the experimental station with ± 0.1 °C temperature stability [1]. With better environmental temperature control, positioning repeatability in a few-micron-level should be achievable for the HXN-Z7-47 robot arm system. This is valuable for the future detector upgrade with high-spatial¬resolution pixel sizes. Figure 1 shows a 3-D model of the HXN-Z7-47 detector robot arm system with the vacuum enclosure for the NSLS-II HXN x-ray microscope. Table 1 summarizes the main design specifications of the HXN-Z7-47 detector robot arm system.

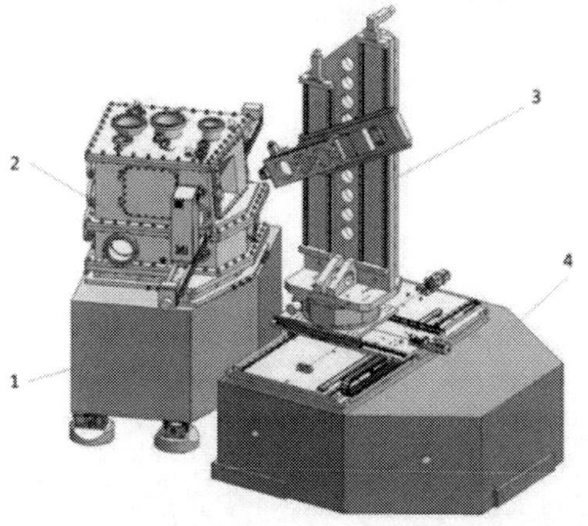

Figure 1. 3D models for the the HXN-Z7-47 detector robot arm system with NSLS-II HXN vacuum enclosure. (1) granite base for HXN vacuum enclosure; (2) HXN vacuum enclosure; (3) detector robot arm; (4) granite base for detector robot arm.

Table 1. Main design specifications for the HXN-Z7-47 detector robot arm system

Parameter	Value	Units
Overall dimension	1700(W) × 2020(L) × 2750(H)	mm
Horizontal diffraction angular range	-5 – +65	deg
Vertical diffraction angular range	-5 – +45	deg
Sample-to-detector distance	0.25 – 0.54	m
Sample-to-detector distance (with base positioning)	0.25 – 2.5	m
Normal load capacity	18	kg
Robot base relocation range	2.5 × 3	m
Unidirectional positioning repeatability	< 20	micron

3. Design of the horizontal base stages with movable granite base

As shown in Figure 1, the HXN-Z7-47 detector robot arm system consists of the following major component groups: a granite base with air-bearing support, a 2-D horizontal base stage, a vertical axis goniometer, a vertical plane robot arm, a 3-D fast scanning stages group, and a 2-D x-ray pixel detector.

The design of the 2-D horizontal base stage is similar to the robot base stage designed for the hard x-ray nanoprobe at the Advanced Photon Source (APS) Sector 26 [4]. Figure 2 shows a 3-D model of the horizontal base stage with a vertical axis goniometer, and a movable granite base. As shown in Figure 2, a THK™ custom-built 2-D horizontal stage (2) and a Huber™ 430 single-circle goniometer (3) are mounted on a granite base (1) to produce a circular motion centered on the sample position in the horizontal plane [4]. There are three air-bearing skids (4) (Airfloat™ AF-1012) integrated with the granite base to provide the robot base station positioning capability. A laser-based positioning tracking system will be applied to ensure the base position accuracy. Table 2 summarizes the design specifications of the 2-D horizontal base stage for HXN-Z7-47 detector robot arm system.

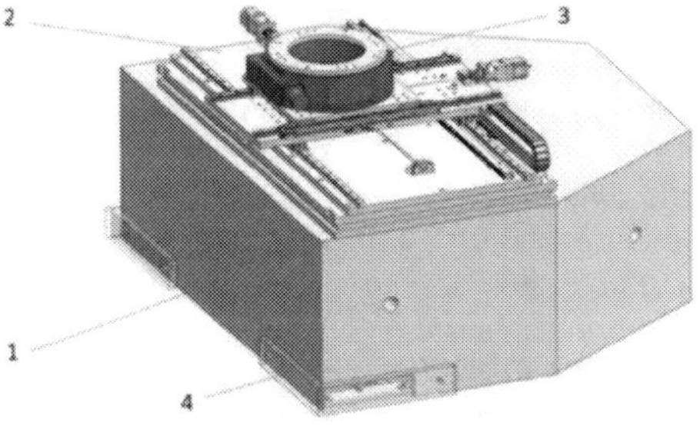

Figure 2. 3-D model of the horizontal base stage with a vertical axis goniometer, and a movable granite base. (1) granite base; (2) 2-D horizontal stage; (3) single-circle goniometer; (4) air-bearing skids.

Table 2. Design specifications for HXN-Z7-47 detector robot arm 2-D horizontal base stages

Parameter	Value	Unit
Overall Dimension	1500 (W) × 750 (L) × 221 (H)	mm
Normal Load Capacity	500	kg
Stage driver type	Stepping motor with planetary gearhead	
Encoder type	Linear grating encoder	
Travel range	1050 (W) × 300 (L)	mm
Min. incremental motion	0.25 – 1.5	micron
Unidirectional repeatability	± 5	micron
Max. speed	600	mm/min

4. Design of the vertical plane robot arm structure

The HXN-Z7-47 detector robot arm system has a unique vertical plane robot arm structure. As shown in Figure 3(left), the vertical plane robot arm structure includes a vertical base (1), two THK™ custom-built vertical linear stages (2) and (3), a robot arm linkage (4), and a 3-D fast scanning stages group (5) to generate the detector distance plus the capability to raster the area detector (6).

Figure 3(right) shows the robot arm linkage sub-assembly. The sub-assembly consists of a linkage base (1), a high-rigidity linear rolling guide (2), and a pair of high-rigidity cross-roller bearings. The upstream bearing (3) is mounted on the linkage base, and the downstream bearing (4) is mounted on the carriage of the rolling guide. The linkage sub-assembly is connected to the carriages of the two vertical linear stages through the pair of cross-roller bearings. The vertical plane robot arm is kinematically positioned in the vertical plane by the two vertical linear stages. A motorized 3-D fast scanning stages group (5) is mounted onto the linkage base to perform the detector raster scan and adjust the detector distance for the 2-D area detector (6). Table 3 summarizes the design specifications of the HXN-Z7-47 detector robot arm vertical linear stages.

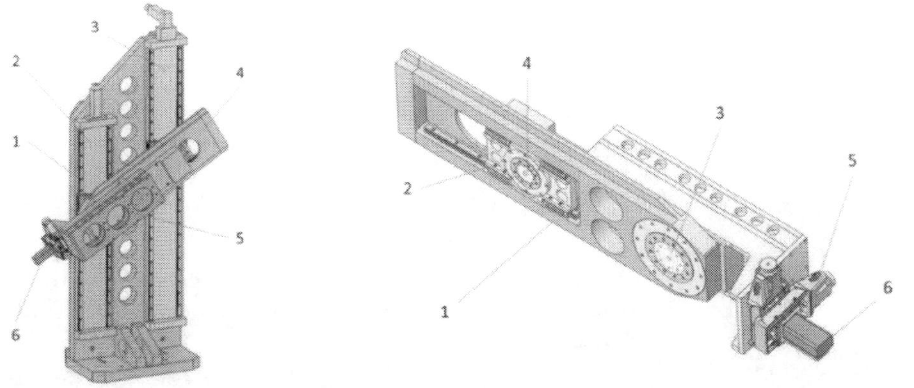

Figure 3. (Left) 3-D models of the HXN-Z7-47 vertical plane robot arm structure: (1) vertical base; (2) and (3) vertical linear stages; (4) robot arm linkage; (5) 3-D fast scanning stages group; (6) 2-D area detector. (Right) 3-D models of the HXN-Z7-47 vertical plane robot arm linkage sub-assembly: (1) linkage base; (2) linear rolling guide; (3) upstream bearing; (4) downstream bearing; (5) 3-D fast scanning stages group; (6) 2-D area detector.

Table 3. Design specifications for HXN-Z7-47 detector robot arm vertical linear stages

Parameter	Value	Unit
Overall Dimension (Upstream stage)	1520 (H) × 200 (W) × 58 (D)	mm
Overall Dimension (Downstream stage)	1788 (H) × 200 (W) × 58 (D)	mm
Travel range (Upstream stage)	850	mm
Travel range (Downstream stage)	1210	mm
Vertical Load Capacity	60	kg
Stage driver type	Stepping motor with planetary gearhead	
Encoder type	Linear grating encoder	
Min. incremental linear motion	0.8/full-step 0.02/micro-step	micron
Min. incremental angular motion	2.1/full-step 0.05/micro-step	micro-rad
Unidirectional repeatability	± 3	micron
Max. speed	160	mm/min

5. Summary

We presented the optomechanical design and specifications of a high-precision detector robot arm system for the scanning x-ray microscope for the HXN beamline at the NSLS-II. Our unique vertical-plane robot arm structure can be adapted to other x-ray microprobes and nanoprobes to enable high-precision x-ray diffraction capability.

References

[1] Preliminary Design Report for Hard X-ray Nanoprobe (HXN) Beamline, 2010 LT-C-XFD-HXN-PDR-001; http://www.bnl.gov/ps/nsls2/beamlines/files/pdf/HXN_pdr.pdf

[2] Huang X, Harder R, Leake S, Clark J and Robinson I, 2012 *J. Appl. Cryst.* **45** 778

[3] Shu D and Chu Y, U.S. Patent application in progress for ANL-IN-13-017

[4] Shu D, Maser J, Holt M, Preissner C, Lai B, Vogt S, Winarski R and Stephenson G B, 2007 *Proc. 21st ASPE Annual Meeting (Monterey, CA)* p123

Acknowledgments

The work at Argonne is supported by the U.S. Department of Energy, Office of Science, under Contract No. DE-AC02-06CH11357. The work at Brookhaven is supported by the U.S. Department of Energy, Office of Science, under Contract No. DE-AC02-98CH10886.

Lithographically-fabricated channel arrays for confocal x-ray fluorescence microscopy and XAFS

Arthur R. Woll[1], David Agyeman-Budu[2], Sanjukta Choudhury[3], Ian Coulthard[4], Adam C. Finnefrock[5], Robert Gordon[6], Emil Hallin[4], and Jennifer Mass[7]

[1] Cornell High Energy Synchrotron Source, Cornell University, Ithaca, NY 14853, USA
[2] Dept. of Materials Science, Cornell University, Ithaca, NY 14853, USA
[3] Geol. Sciences, Univ. of Saskatchewan, 114 Science Place, Saskatoon SK S7N 5E2 Canada
[4] Canadian Light Source, 44 Innovation Blvd, Saskatoon, SK S7N 2V3, Canada
[5] Merck Research Laboratories, 770 Sumneytown Pike, West Point, PA 19486, USA
[6] PNCSRF, APS Sector 20, 9700 S. Cass Ave. 435E, Argonne, IL 60439 USA
[7] Winterthur Museum, Winterthur, DE, 19735, USA

E-mail: `arthurwoll@cornell.edu`

Abstract. Confocal X-ray Fluorescence Microscopy (CXRF) employs overlapping focal regions of two x-ray optics—a condenser and collector—to directly probe a 3D volume. The minimum-achievable size of this probe volume is limited by the collector, for which polycapillaries are generally the optic of choice. Recently, we demonstrated an alternative collection optic for CXRF, consisting of an array of micron-scale collimating channels, etched in silicon, and arranged like spokes of a wheel directed towards a single source position. The optic, while successful, had a working distance of only 0.2 mm and exhibited relatively low total collection efficiency, limiting its practical application. Here, we describe a new design in which the collimating channels are formed by a staggered array of pillars whose side-walls taper away from the channel axis. This approach improves both collection efficiency and working distance, while maintaining excellent spatial resolution. We illustrate these improvements with confocal XRF data obtained at the Cornell High Energy Synchrotron Source (CHESS) and the Advanced Photon Source (APS) beamline 20-ID-B.

1. Introduction

Over the past 10 years, confocal x-ray fluorescence microscopy (CXRF) has become an increasingly common approach to the problem of obtaining elemental composition distributions in three dimensions. As described in several recent reviews [1], it has been demonstrated at numerous synchrotrons worldwide as well as with lab-based sources, and applied to a wide variety of applications. Recently, we demonstrated the first use of lithographically-etched channels in silicon as a collection optic for CXRF [2]. In this approach, the collection optic determines one of the three dimensions of the confocal probe volume, whereas the remaining two are defined by the incident beam. Ref. [2] demonstrates an order-of-magnitude improvement in resolution when compared with state-of-the-art demonstrations of CXRF, virtually all of which use polycapillaries for collection. However, the optics employed in Ref. [2] have limited practical appeal, exhibiting relatively poor collection efficiency compared to polycapillaries and a very small working distance, approximately 0.2 mm. The collection efficiency of these optics,

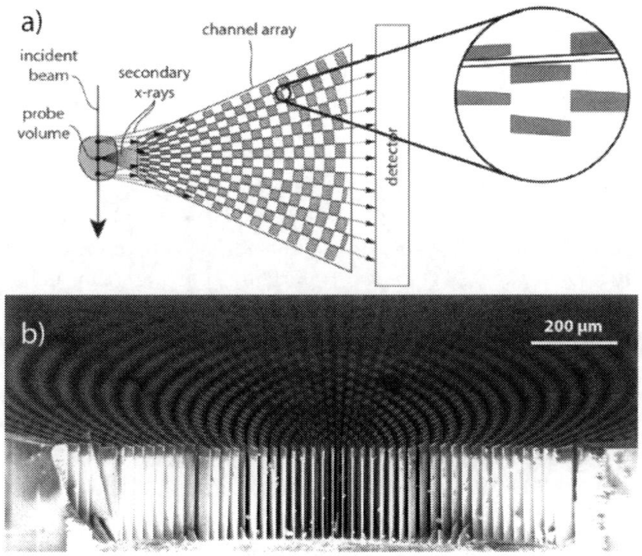

Figure 1. (a) Schematic illustration of CXRF configuration using collimating channels that are formed from a set of staggered, absorbing pillars. The inset illustrates that the pillar edges are tapered away from the axis of each channel to absorb reflected photons. (b) Scanning electron micrograph of a channel array formed by an arrangement of staggered, tapered pillars etched into a silicon substrate. The optic shown has an effective channel width of $\approx$2.2 μm, but the effective channel depth is over 200 μm.

defined [2] as the fraction of fluorescent photons emitted from a point source captured and transmitted by the optic, was limited by the channel depth, which varied from 30-50 μm for channel widths of 1-5 μm. The small working distance of these optics was necessitated by the large, diffuse reflectivity of the channels. As a result, the collimating power or angular selectivity δ of the channels was much larger than the designed value, w/l, the ratio of the channel width w and length l.[1] Because the spatial resolution r of the channel array is approximately $w + \delta f$, (where f is the working distance from the sample to optic), large δ requires small f in order to achieve small values of r.

Here, we report fabrication of an improved channel array design, illustrated in figure 1a, that addresses both of these limitations. A key feature of this approach is that channels are effectively formed by a series of staggered pillars. This arrangement increases the distance between features without increasing the channel width, allowing much greater etch depth. Figure 1b shows a scanning electron micrograph (SEM) of an optic fabricated with such a design, illustrating effective channel depths upwards of 200 μm. The optic in figure 1b was designed with 2 μm channels, so that the effective aspect ratio of channels is approximately 100:1. The second key feature of this design, indicated in the inset of figure 1a, is that the pillars forming each channel are tapered away from the channel axis. Such tapering successfully reduces the angular acceptance of the channels, thus allowing an increase in working distance while maintaing good spatial resolution. Below, we present an overview of recent results obtained at both CHESS and APS 20-ID-B using channel arrays that incorporate these ideas.

[1] Note that this definition of δ is a factor of 2 smaller than that defined in Ref. [2], resulting in a compensating change in the definition of channel resolution r.

Journal of Physics: Conference Series **493** (2014) 012028 doi:10.1088/1742-6596/493/1/012028

2. Results

Based on the principles described above, new optics were designed, fabricated and characterized using techniques described in Ref. [2]. The angular selectivity of the channels is measured by directing a well-collimated incident beam through a single channel, then rotating the optic about its focal point and measuring the transmitted intensity. Two optics representing the new design, having channel dimensions of 1 μm $\times$ 4 mm and 7 μm $\times$ 4 mm, were found to have angular selectivities δ of 2.2 and 3.3 mrad, respectively, at 10 keV. These values represent a factor-of-two or better reduction compared to the straight channels described in Ref. [2], and are sufficient to allow modest increases in f without greatly increasing r, as shown below. Nevertheless, they are significantly larger than their ideal, geometric collimation w/l, 0.25 and 1.75 mrad. Comparison with ray-tracing simulations suggest that the broadening of δ for these channels is induced by refraction at the front edges of the tapered pillars (see Fig. 1a).

We have employed these optics for confocal XRF at both the G1 station at CHESS and APS station 20 ID B, for both XRF mapping and confocal XAFS [3, 4]. Tests at CHESS employing the 7 μm $\times$ 4 mm channel array demonstrate that the increased channel depth of these arrays gives rise to a commensurate increase in collection efficiency. To measure this efficiency, we direct a focused incident x-ray beam, approximately 3μm in diameter, onto a metal film, approximately 200-nm thick, such that the illuminated area of the film acts as a point source of x-ray fluorescence. By comparing the fluorescence intensity from this source measured with and without our optic in place, the collection efficiency may be determined as a fraction of 4π. For this array, we obtain a value of 0.14% near 9 keV – a factor of 10 greater than that reported previously, and nearly equivalent to that of a polycapillary measured with an otherwise identical setup.

An additional, critical test of the new design is whether comparable resolution to that reported in Ref. [2] could be achieved with a larger working distance f. This test was carried out at APS 20-ID-B, for which a KB mirror pair was used to focus the incident beam to approximately 1.5 μm (HZ) $\times$ 2μm (VT). The 1 μm $\times$4 mm channel array, with a working distance of 0.5 mm, was employed. The depth resolution d_R [2], was measured by scanning the surface of NIST SRM 1834 through the confocal volume and fitting to the error function, modified to include attenuation [5]. An example of such a scan and fit is shown as the inset to figure 2. The depth resolution corresponds to the sharpness of the left-hand edge of this curve. Figure 2 shows the values of d_R obtained for different emission lines in the sample. The values are all close to 1.7 μm, rising to just above 2 μm below 4 keV.

Figure 3 shows a virtual 2D composition map, obtained with confocal XRF at CHESS, of the near-surface region of a grain of dry brown rice [6]. In this case, the 7 μm $\times$ 4 mm channel array was used. Due to its large variety of elemental constituents, the sample serves to illustrate excellent spatial resolution ($d_R = 5$ μm in this case) for low energy emission. The figure shows that the outer surface of the rice grain, or bran, is composed of two layers with distinct composition. Both layers contain potassium, but manganese and phosphorous are selectively distributed in the upper and lower layers, respectively. Since the depth resolution of state-of-the-art, but traditional CXRF implementations generally exceeds 20 μm below 4 keV [2], the potassium and phosphorous concentrations in figure 3 would be severely obscured in CXRF maps of this sample obtained with polycapillaries as the collection optic.

3. Summary

In this article, we have shown that lithographically-defined microchannel arrays constitute a practical and advantageous collection optic for confocal x-ray fluorescence microscopy. Although not shown here, we have also employed these optics for XANES in confocal mode. We anticipate that these optics will significantly enhance and broaden the scope of application of confocal XRF, particularly for light elements and samples with spatial variation on the 1 μm scale.

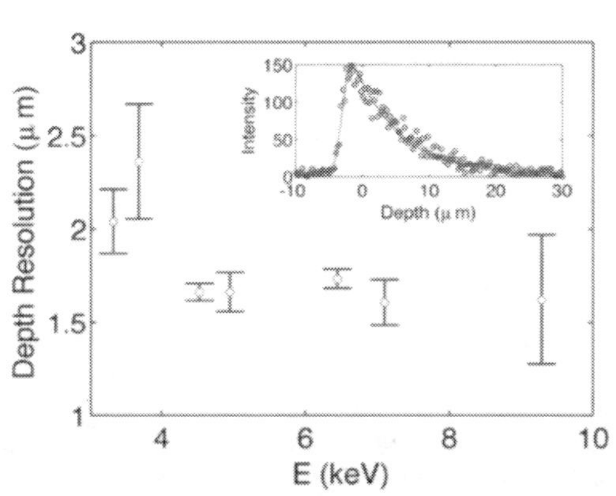

Figure 2. (Color online) Depth resolution $d_R(E)$ for a confocal XRF arrangement employing a channel array with 1 μm× 4 mm nominal channel dimensions. The inset shows the intensity distribution and fit for potassium Kα, one of the curves employed to obtain $d_R(E)$.

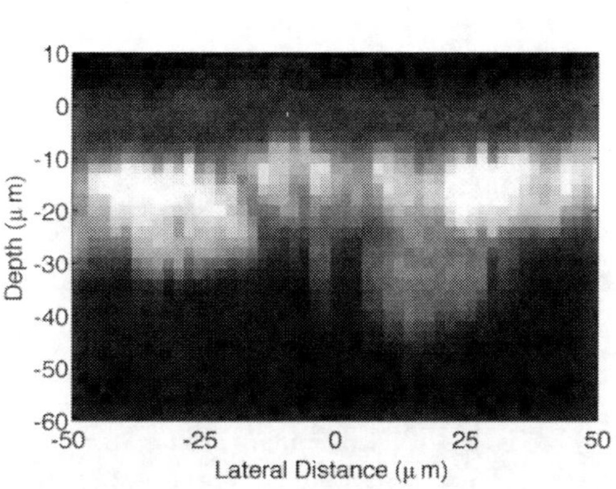

Figure 3. (Color online) Representation of a 2D confocal XRF scan of the the near-surface region of a grain of brown rice. In the image, the Kα intensities of potassium, phosphorus, and manganese are normalized to their maximum, then encoded as red, green, and blue, respectively.

Acknowledgements

This work was supported by the Cornell Center for Materials Research, a National Science Foundation (NSF) Materials Research Science and Engineering Center (NSF-DMR-0520404). This work was performed in part at NSF-supported members of the National Nanotechnology Infrastructure Network, the Lurie Nanofabrication Facility and the Cornell NanoScale Facility(Grant ECCS-0335765). This work was performed in part at the Cornell High Energy Synchrotron Source, supported by the NSF and NIH-NIGMS (NSF-DMR-0225180). PNC/XSD facilities at the Advanced Photon Source, and research at these facilities, are supported by the U.S. Department of Energy - Basic Energy Sciences, a Major Resources Support grant from NSERC, the University of Washington, the Canadian Light Source and the Advanced Photon Source. Use of the Advanced Photon Source, an Office of Science User Facility operated for the U.S. Department of Energy (DOE) Office of Science by Argonne National Laboratory, was supported by the U.S. DOE under Contract No. DE-AC02-06CH11357. JM and ACF acknowledge the Lenfest Foundation for additional support.

References

[1] Kanngießer B, Malzer W, Mantouvalou I, Sokaras D and Karydas A 2011 *Appl. Phys. A: Mat. Sci. and Proc.* 1–14; Perez R D, Sanchez H J, Perez C A and Rubio M 2010 *Radiat. Phys. Chem.* **79** 195–200; Fittschen U E A and Falkenberg G 2011 *Anal. Bioanal. Chem.* **400** 1743–1750

[2] Woll A R, Agyeman-Budu D, Bilderback D H, Dale D, Kazimirov A Y, Pfeifer M, Plautz T, Szebenyi T and Untracht G 2012 *SPIE Optics and Photonics 2012* vol 8502 ed Goto S, Morawe C and Khounsary A M (SPIE) pp 85020K–1 – 85020K–14

[3] Silversmit G, Vekemans B, Appel K, Schmitz S, Schoonjans T, Brenker F E, Kaminsky F and Vincze L 2011 *Anal. Chem.* **83** 6294–6299

[4] Luhl L, Mantouvalou I, Malzer W, Schaumann I, Vogt C, Hahn O and Kanngiesser B 2012 *Anal. Chem.* **84** 1907–1914

[5] Mantouvalou I, Malzer W, Schaumann I, Luhl L, Dargel R, Vogt C and Kanngiesser B 2008 *Anal. Chem.* **80** 819–826

[6] Mihucz V G, Silversmit G, Szalóki I, de Samber B, Schoonjans T, Tatár E, Vincze L, Virág I, Yao J and Záray G 2010 *Food Chemistry* **121** 290 – 297

Radiation Damage of Myoglobin Crystals in Weak Stationary Electric and Magnetic Fields

C B Trame[1,2], **M Dragovic**[2], **H-J Chiu**[1,2]
[1]Joint Center for Structural Genomics (http:/www.jcsg.org), USA
[2]Stanford Synchrotron Radiation Lightsource, SLAC National Accelerator Laboratory, Menlo Park, CA 94025, USA

E-mail: cbtrame@slac.stanford.edu

Abstract. Radiation damage is one of the bottlenecks in the field of structural biology. Cryo-cooling of protein crystals provided a breakthrough in the 1980s and resulted in significant reductions in radiation damage. Other factors positively influencing the progression of damage include the application of radical scavengers and reductions in the experimental beam size. Here we study the impact on radiation damage of applying static magnetic and electric fields during protein diffraction experiments, ultimately probing the Lorenz force effect on primary photoelectrons and secondary Auger electrons, which both contribute to the damage process. The design of a special mounting pin using graphene for applying electric fields on a crystalline sample is described. Analyses of myoglobin protein crystals exposed to the fields of ~40 mT and -300 V show a slower global radiation damage rate and also changes in the progression of specific damage process on the molecular level, in particular at doses extending beyond the Garman limit of 30 MGy.

1. Introduction

Myoglobin is one of the most studied biological molecules, especially in regard to radiation damage using X- and γ-rays [1]. The consequences of exposure to high energy photons on protein crystals, are specific structural changes, which lead to changes in global crystallographic parameters ($I/\sigma(I)$, R_{cryst}, mosaicity, anomalous signal and lattice dimensions). The structural changes are sensitive to chemical environment and are influenced by the distribution of photo- and Auger-electron cascades released in the photo-absorption process. It is the charge properties and the distribution of the resulting electrons and ions within protein crystals that prompted us to study the radiation damage effect in electric and magnetic fields. In addition, electric and magnetic fields are also known to influence protein crystallization process [2]. Qualitative crystallographic analyses of myoglobin crystals mounted on a newly designed pin using graphene support compared to crystals mounted on standard nylon loops suggests an improved crystal mount and data quality for protein diffraction experiments.

---

[1] To whom any correspondence should be addressed: cbtrame@slac.stanford.edu

2. Experiment

A special HV (High Voltage) pin was designed in order to apply electrostatic fields on protein crystals during diffraction experiments. Figure 1 shows the scheme of the design, which allows a free phi-rotation, while keeping a constant bias voltage on the crystal. For the application of magnetic fields, a rare earth permanent bar magnet (N52 type) was placed ~3 cm away from the sample, with its N or S pole oriented at ~35° to the phi axis. The stationary magnetic field was determined to be ~0.04 T around the crystal mounting graphene surface using 410 Lake Shore Gaussmeter, and the value varied by ~2% due to the background magnetic fields from other mounting components (3, 5 and 6).

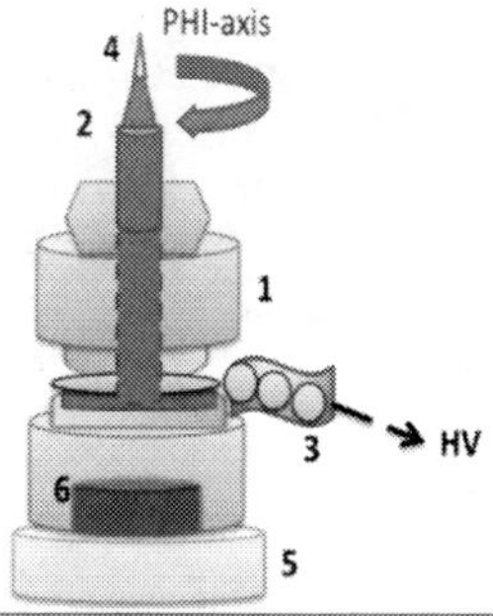

Figure 1. HV pin consisting of insulating plastic body (1) and a magnetic stainless steel core (2) glued to a thin disk, which is attracted to a flexible chain of magnetic beads inserted in a loose shrinking tube (3). The pin core and the tip with the graphene foil (4) are connected via low ohm conducting beads to the HV power supply (Model Ortec556). The pin is positioned on the goniometer base (5) via permanent magnet (6) and is centered in the middle of the 100K cold LN2 stream.

Sperm whale myoglobin was crystallized according to previously published methods [3]. All experiments were performed on the SSRL bending magnet beamline 14-1. Multiple data sets were collected from six myoglobin crystals (Table 1). All crystals diffracted to ~1.00-1.25 Å. Typically 10 data sets in the same 120° phi range for each crystal were collected at 1.0 Å wavelength and at 100 K. All data sets were processed using identical XDS data processing parameters. For the purpose of comparing radiation damage under different parameters, we chose to monitor 1.9-2.0 Å resolution shell. The crystal sizes were determined using a Navitar Lens based microscope camera. The size of each loop mounted crystal was further verified using Fe fluorescence signal integrated around Fe K-edge with the excitation wavelength of 1.0 Å. An example of fluorescence profile used to determine crystal size is shown in Figure 2.

Table 1: List of myoglobin crystals used in this study

Crystal ID.	Pin type	Mount	Operation sequence[a]
X23	HV	Graphene	HV -300V / B(N) / B(S)[c]
X24	HV	Graphene	B(S)
X25	Cu[b]	Nylon loop	--
X30	Cu[b]	Nylon loop	--
X33	HV	Graphene	HV -300V / B(S)
X34	Cu[b]	Nylon loop	--

[a] In the cases that both electric field and magnetic fields were applied during the data collection, the electric field was applied first and followed by magnetic field. B(N) and B(S) mean that the north or south poles face sample respectively.
[b] Copper pin from Hampton Research.
[c] The magnetic polarity B(N) was changed to B(S) after 3rd data set.

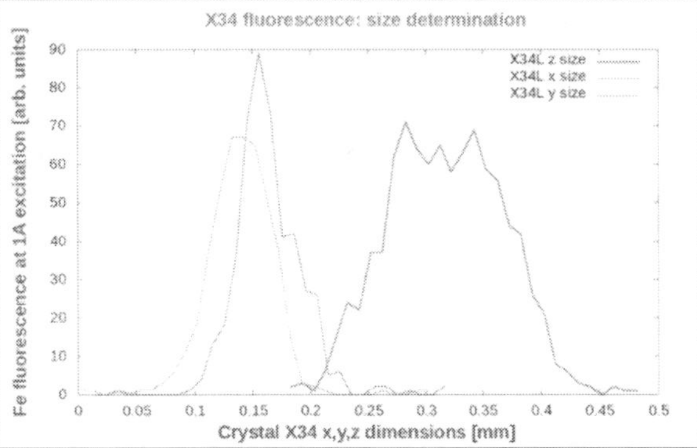

Figure 2. Measured dimensions of myoglobin crystal X34 in (x,y,z) as a profile of an integrated Fe fluorescence signal using a VORTEX Si-drift detector. The overall dimensions of ~ 175x101x93 μm^3 match well with the size determined by a microscope camera. Signal to noise (loop+crystal/loop) ratio of this method is 50-60 with the beam sizes of (30x30)μm.

With a photon flux of 1.02×10^{11} photons/sec at 1.0 Å wavelength, the program RADDOSE [4] calculated a radiation dose of ~4.5 MGy per data set. Typically 10 data sets were collected until a ~10-fold fall-off in I/σ(I) in the 1.9-2.0 Å resolution shell was reached. The experimental dose limit of 30 MGy was reached during the ~6th - 7th data set for each crystal. Due to the variation in photon flux and crystal size in different experiments, the scaling factor S, proportional to X-ray dose, was calculated using the general formula [5] and was applied to each data set:

$$S \propto (I_o/\lambda V)[1-\exp(-\mu_{abs} t)]$$

Here S is proportional to the absorbed energy per unit mass by a crystal with thickness t and volume V. The absorption coefficient μ_{abs} is calculated for an incident wavelength λ and beam intensity of I_o. That crystal dependent dose scaling factors S, ranging between 2 and 3.5, were then multiplied by the data set number N for each of the crystals, and used as an abscissa in figure 3.

3. Results and Discussion

The XDS data reduction resulted in the fitted crystallographic parameters indicating the radiation damage progression for each of the crystals, as shown in figure 3.

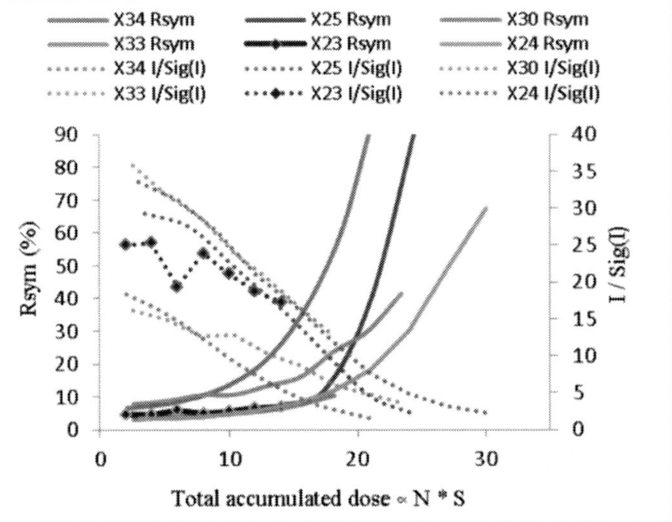

Figure 3. The increase of Rsym and the decrease of I/σ(I) as a function of accumulated X-ray dose show that the data quality deteriorates faster in crystals X25, X30 and X34, which were mounted in standard nylon loops, as compared to crystals X23 and X33, which were mounted on graphene foil and collected in the presence of magnetic and electric fields.

Our preliminary results suggest that myoglobin crystals suffer radiation damage at a slower rate when both electric and magnetic fields are applied simultaneously. Mounting crystals on a graphene surface also has the advantage of reducing radiation damage as compared to Hampton-style nylon loops. Specific structural damage at the atomic level reveals itself through density differences around the heme prosthetic group and charged residues, when comparing for example electron density maps of the 8th set of X33 and X34 (Figure 4a and 4b).

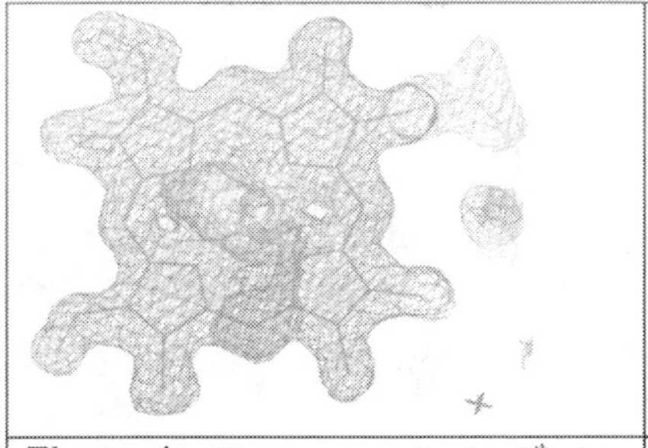

Figure 4a. Heme group in the 8th data set of X34. The 2Fo-Fc map is colored in magenta and contoured at 1.1σ. The Fo-Fc maps are shown in green and red and are contoured at ±3.5σ. There is more positive Fo-Fc density spread out around the heme plane in X34 than in X33.

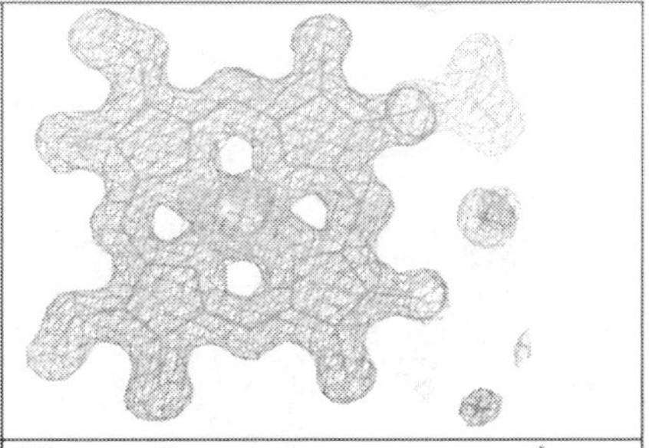

Figure 4b. Heme group in the 8th data set of X33. The 2Fo-Fc map is colored in blue and contoured at 1.1σ. The Fo-Fc maps are shown in green and red and are contoured at ±3.5σ. The positive Fo-Fc density is centrally localized around Fe atom. The Fe atom stays in the heme plane.

The reduced amount of solvent/water due to surface tension effects for graphene foils over nylon loops may be a contributing factor in reducing radiation damage progression. The fact that the observed dose dependent effects are small at the beginning and become more apparent towards the end of data collection is possibly due to limited range of electromagnetic fields that are available in our current experimental setup. For future experiments, the hardware setup will be improved to allow higher electric and magnetic fields to be applied to the samples. It appears, that freezing myoglobin crystals in the presence of a magnetic field has a negative effect on the crystal, which tends to split. Similar tendency in worsening the diffraction pattern was observed with the application of +300V field in combination with B fields (hence data not shown). A fluorescence size profile scan was proved to be a useful tool for measuring the crystal size.

Acknowledgments

The JCSG is funded by NIGMS/PSI U54 GM094586. SSRL operations are funded by DOE BES, and the SSRL SMB program by DOE BER, NIH NCRR BTP and NIH NIGMS. Biochemistry was supported by NIH Grant R21DC011332. Authors thank Mathews Irimpan for providing myoglobin crystals and Abhinav Kumar for help with graphics.

References

[1] Hersleth H-P, Hsiao Y-W, Ryde U, Görbitz C H, Andersson K K 2008 *Chemistry & Biodiversity* **5**: 2067
[2] Sazaki G, Moreno A and Nakajima K 2004 *J. Cryst. Growth* 262: 499
[3] Kondrashov D A, Zhang W, Aranda R, Stec B and Phillips G N 2008 *Proteins* **70**: 353
[4] Murray J W, Garman E F and Ravelli R B G 2004 *J. Appl. Cryst.* **37**: 513
[5] Paithankar K S, Owen R L and Garman E F 2009 *J. Synch. Rad.* **16**: 152

Completion of the Brightness Upgrade of the ALS*

C Steier[1], A Madur[1], B Bailey[1], K Berg[1], A Biocca[1], A Black[1], P Casey[1], D Colomb[1], B Gunion[1], N Li[1], S Marks[1], H Nishimura[1], C Pappas[1], K Petermann[1], G Portmann[1], S Prestemon[1], A Rawlins[1], D Robin[1], S Rossi[1], T Scarvie[1], R Schlueter[1], C Sun[1], H Tarawneh[1], W Wan[1], E Williams[1], L Yin[2], Q Zhou[2], J Jin[2], J Zhang[2], C Chen[2], Y Wen[2] and J Wu[2]

[1] LBNL, Berkeley, CA 94720, USA
[2] SINAP, Shanghai, China

E-mail: `CSteier@lbl.gov`

Abstract. The Advanced Light Source (ALS) at Berkeley Lab remains one of the brightest sources for soft x-rays worldwide. A multiyear upgrade of the ALS is underway, which includes new and replacement x-ray beamlines, a replacement of many of the original insertion devices and many upgrades to the accelerator. The accelerator upgrade that affects the ALS performance most directly is the ALS brightness upgrade [1], which reduces the horizontal emittance from 6.3 to 2.0 nm (2.5 nm effective). Magnets for this upgrade were installed in late 2012 and early 2013 followed by user operation with the reduced emittance.

1. Introduction

The ALS produces light over a wide spectral range for users from far infrared (IR) to hard x-rays with the core spectral region being soft x-rays. In this core region (relevant to life-science, chemistry, catalysis, surface science, nano-science, and complex materials), the ALS remains competitive with the newest synchrotron radiation sources worldwide (see brightness comparison in Fig. 1).

2. Brightness Upgrade Lattice Design

The ALS lattice has a triple bend achromat structure, with a fixed, large defocusing gradient in the bending magnets. Originally, there were only 2 families of sextupoles, with 4 sextupole magnets in each arc. An attractive set of possible upgrade lattices was found with higher straight section dispersion and an integer tune two units higher than the old lattice [2]. Those lattices have natural emittances of just above 2 nm (compared to the more than 6 nm of the old lattice). Later on, more systematic techniques [1, 3, 4] were used to find the global optimal lattices in terms of emittance or brightness. In those studies an additional family of low emittance lattices was found with very small horizontal beta function (order of 0.5 m) in the straights at much higher phase advance, which increase the brightness by better matching to the photon diffraction ellipse. Such low-beta insertions are still under investigation for selected straights.

*The Advanced Light Source is supported by the Director, Office of Science, Office of Basic Energy Sciences, of the U.S. Department of Energy under Contract No. DE-AC02-05CH11231.

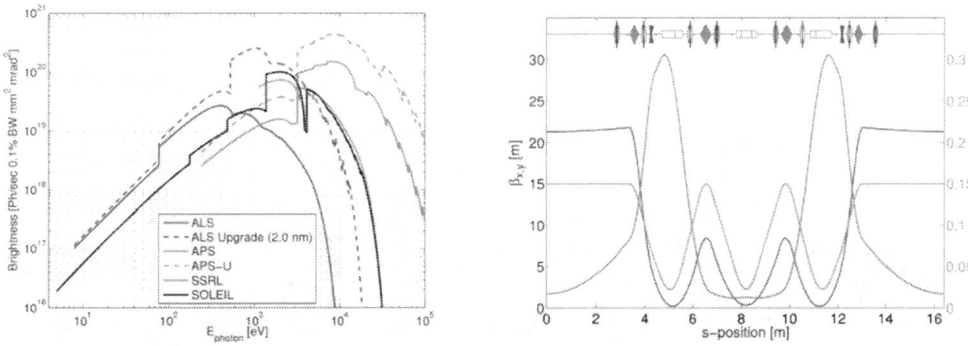

Figure 1. Left: Comparison of ALS brightness before and after the brightess upgrade with several other existing light sources. Right: Lattice functions of ALS upgrade lattice with 2.0 nm natural emittance (1.9 GeV).

The high-beta lattices are within the range of the existing quadrupole magnets. However, the original sextupoles were too weak and the dynamic aperture would have been very poor. Both challenges were overcome with the addition of sextupoles in the straight sections.

With the addition of new magnets and the changed strength of existing interlocked magnets, parts of the safety analysis for top-off operation needed to be redone. The analysis was completed in time for the installation shutdown and no hardware changes were necessary.

3. Magnet Design and Production

The design [5] of the new sextupoles was performed in a collaboration by LBNL and SINAP and was finished in 2011. Because of space constraints, 3 different sextupole magnet designs are used. One of the families is optimized for small hysteresis and fast time response and has a closed yoke. It is also used as primary correctors in the fast orbit feedback. All new sextupoles also contain skew quadrupole coils (half of them are currently connected to power supplies). This allows to improve the vertical beamsize stability in the ALS by providing an effective correction of the small but relevant skew quadrupole errors of the planar insertion devices.

Magnet production, carried out at SINAP, started with prototype magnets just after the design reviews in early 2011. The pole shapes were manufactured by wire-edm on fully assembled magnet cores to achieve excellent field quality. Manufacturing was completed in summer 2012, on time to achieve the project installation milestones. During construction there was a detailed quality assurance program and all magnets were fully qualified by electrical, mechanical, and magnetic measurements. Precise fiducialization was carried out both mechanically and with the help of magnetic measurements. All magnets exceeded the necessary field quality requirements.

4. Installation

In order to create sufficient space in all locations where new magnets were going to be installed, several modifications of vacuum chambers and stands were completed in 2012. The installation started late in 2012 during short maintenance shutdowns, with 13 of the new sextupoles being installed ahead of the main installation shutdown. This allowed to test their corrector functionality (time response, hysteresis) and to incorporate them into slow and fast orbit feedback. The remainder of the 48 magnets were installed during the 2013 spring shutdown (see photos in Fig. 2). At the same time, all new power supplies and equipment protection systems were installed, the topoff interlock ranges enlarged and the interlocks retested.

Figure 2. Left: SHD magnet installed between two of the QF and QD quadrupoles. Right: Ribbon cutting celebration after successful installation.

5. Commissioning

Migration to the new lattices was quick (few hours), after all magnet polarities and magnet transfer functions had been verified in a beam based way in the old lattice. Simulations had predicted excellent dynamic and momentum aperture as well as lifetime for the optimized upgrade lattices [1, 7]. These predictions were quickly confirmed. Further commissioning included optimizing the harmonic sextupole settings, updating the ID feed-forward algorithms (tune, beta beating, coupling) for the new lattice, implementing the new dispersion bump [6] (see Fig. 3) for the fs-slicing facility and retesting the top-off interlocks with new ranges.

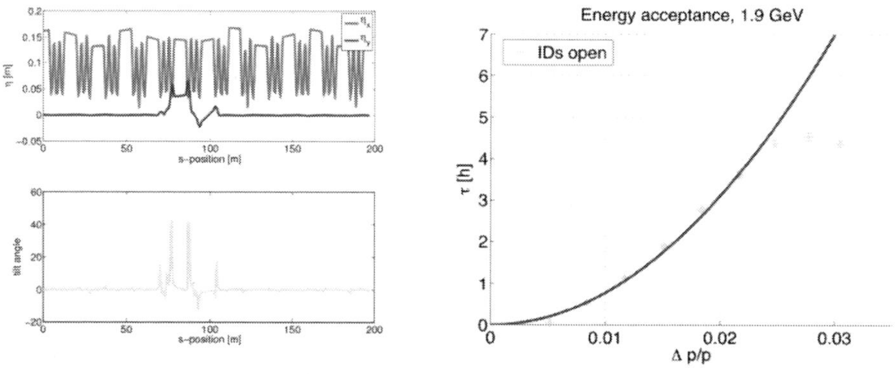

Figure 3. Left: New fs-slicing optics that provides the spatial separation of the energy sliced beam in the new upgrade lattices. Right: Lifetime vs. RF amplitude scan for the upgrade lattice confirming improved momentum acceptance compared to operation before the upgrade.

The dispersion bump was refined after final lattice optimizations. The dynamic aperture and momentum aperture (see Fig. 3) including the fs-slicing lattice insertion are similar to the bare lattice results and commissioning continued on a fast pace. The new lattices also provide a larger, intrinsic horizontal separation of the sliced electron beam. We are currently in the process of evaluating how to make best use of this and expect that it will eventually allow a much better signal to noise ratio for the slicing facility.

Optimizing the photon beamlines for the new beam dimensions progressed quickly and user beamlines were able to resolve the brightness increase (see Fig. 4). The Touschek beam lifetime after the upgrade, despite the smaller horizontal and slightly smaller vertical emittance is larger than before the upgrade, due to the larger dynamic momentum aperture, as predicted.

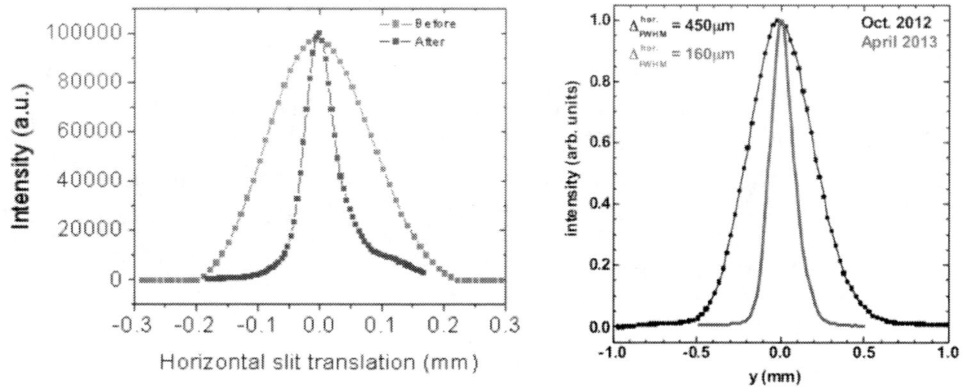

Figure 4. Comparison of the horizontal beam profile before and after the upgrade measured at beamlines 12.3.2 and 6.3.1 showing the factor of three improvement in brightness (vertical scale is renormalized, both beamlines have different magnification factors).

6. Summary

An upgrade project has been completed improving the brightness of the ALS by reducing the horizontal emittance from 6.3 to 2.0 nm. This resulted in a brightness increase by a factor of three for bend magnet beamlines and at least a factor of two for insertion device beamlines. The ALS now has one of the smallest horizontal emittances of all operating 3rd generation light sources. Initial user operations has been very successful. Most beamlines have been able to benefit significantly from the upgrade. No interruptions during the first months of user operations were related to the upgrade.

7. Acknowledgements

The authors want to thank ALS management, as well as the ALS user community for support to develop the ALS storage ring further, as well as their support in securing funding for this upgrade project. We are very grateful to D. Blough who helped us with Quality Assurance. We thank M. Kunz, N. Tamura, and K. Jenkins for the beam profile measurements.

References

[1] C. Steier, et al., NIM A, DOI: 10.1016/j.nima.2010.11.077
[2] H. Nishimura, et al., PAC 2007, Albuquerque, p. 1970.
[3] D. Robin, et al., Phys. Rev. STAB 024002 (2008).
[4] L. Yang et. al, Nucl. Instr. Meth. A 609 (2009) 50-57.
[5] A. Madur, et al., IEEE Trans. Appl. Superc. 22
[6] C. Sun, et al., New fs-slicing optics, IPAC2012
[7] C. Sun, et al., Phys. Rev. STAB 15, 054001 (2012).

The Latest Status of NSLS-II Insertion Devices

Toshi Tanabe, Charles Kitegi, Ping He, Marco Musardo, Oleg Chubar James Rank, Peter Cappadoro, Huston Fernandes, David Harder and Todd Corwin

Brookhaven National Laboratory, Photon Science Directorate, Upton, NY, USA

ttanabe@bnl.gov

Abstract. The National Synchrotron Light Source-II (NSLS-II) project is now in the final stage of construction. The Linac, the Booster synchrotron, and the Storage Ring magnets girder assemblies have been installed. The first damping wiggler has been delivered and its field characteristics are carefully measured. A Three Pole Wiggler (3PW) and Apple-II type elliptically polarizing undulators (EPUs) have been fabricated by the vendors. Two 3.0m long in-vacuum undulators (IVUs) and one 1.5m long IVU are almost complete and waiting for factory acceptance tests. One 3.0m long IVU for Inelastic X-ray Scattering beamline is in fabrication by a different vendor. Recently two 2.8m long IVUs for long straight sections (LSSs) have been added to the project for "future beamlines". In addition, two 1.5m long IVUs and one 2.8m long IVU for LSSs have been procured for Advanced Beamlines for Biological Investigations with X-rays (ABBIX) project funded by National Institure of Health (NIH). Further, two 3.5m long EPUs for LSSs are being designed for NSLS-II Experimental Tools (NEXT) -Major Item of Equipment (MIE) project. To succeed these conventional IVUs, PrFeB based cryo-permanent magnet undulator (CPMU) is considered as next generation device of hard X-ray sources. An In-Vacuum Magnetic Measurement System (IVMMS) for cold *in-situ* Hall probe mapping of CPMUs up to 1.5m in length has been developed. Summary of the current status of each project and future plans for the NSLS-II ring will be discussed.

1. Introduction

The Photon Sciences Directorate at Brookhaven National Laboratory is constructing the National Synchrotron Light Source II (NSLS-II) facility. NSLS-II will be able to accommodate at least 58 beamlines for scientific programs, with additional beamlines possible through canted insertion devices and multiple hutches. There are two avenues of beamline development: Type I – in which NSLS-II is responsible for securing the funding, designing and constructing the beamline; and Type II – for which a group external to NSLS-II is funding, designing and constructing the beamline. Currently funded insertion device beamlines are (6 +2 branches) NSLS-II project beamlines, three NIH-funded ABBIX beamlines, six NEXT-MIE project beamlines and two Type-II beamlines by NIST and the other by Columbia University. All the acronyms for the different NSLS-II beamlines can be found in ref [1].

2. List of Insertion Devices

Table I shows a list of NSLS-II project insertion devices. Two 2.8m long U23s have been added recently to equip the facility for future beamlines. The insertion devices used for NEXT-MIE are

Table 1: NSLS-II Project Baseline Insertion Devices

	U20	**U21 (SRX)**	**U22 (IXS)**	**U23L**	**EPU49 (CSX)**	**DW**	**3PW**
Type	IVU	IVU	IVU	IVU	Apple-II	Hybrid PM	Hybrid PM
Length [m]	3.0	1.5	2x 3.0	2.8	2 x 2.0	2 x 3.4	0.3
Period Length [mm]	20.0	21.0	22.0	23,0	49.0	100.0	
Magnetic Gap [mm]	5.2	6.4	7.5	6.1	11.5	15.0	28.0
B_{peak}[T]	1.0	0.9	0.78	0.95	0.94 (lin) / 0.57 (heli) 0.72 (vlin) / 0.41 (45°)	1.8	1.13
Maximum K	1.8	1.8	1.5	2.05	4.3 (lin) / 2.6 (heli) 3.2 (vlin) / 1.8(45°)	15.2	
Fundamental Photon Energy	1830	1560	1800		180 (lin) / 230 (heli) 285 (vlin) / 400(45°)		
Ec [KeV]						10.7	6.8
Max Total Power [kW]	8.0	3.6	4.5 x 2	7.2	8.8 (lin)	64	0.34

Table 2: NEXT-MIE Insertion Devices

	EPU105 (ESM)	**EPU49 (ESM)**	**EPU57 (SIX)**
Type	Apple-II	Apple-II	Apple-II
Length [m]	2.8	1.4	2 x 3.5
Period Length [mm]	105.0	49.0	57.0
Min. Magnetic Gap [mm]	16.0	16.0	16.0
B_{peak}[T]	1.14 (lin) /0.74 (vlin)	0.72 (lin) /0.47 (vlin)	1.14 (lin) /0.74 (vlin)
Maximum K	11.2 (lin) / 7.23* (heli) 7.23* (vlin)	3.27 (lin) / 2.53 (heli) 2.14 (vlin)	4.41 (lin) / 3.55 (heli) 3.06 (vlin)
Fundamental Photon Energy	12.8 (lin) / 15.3* (heli) 30* (vlin)	275 (lin) / 236 (heli) 530 (vlin)	140 (lin) / 110 (heli) 264 (vlin)
Max Total Power [kW]	10.1 (lin)	2.0 (lin)	2x7.0 (lin)

* For EPU105, Keff values for Circular (C) and Linear Vertical (LV) shown here are the maximum values permitted owing to the power loading issues on Bend Magnets and/or ID chamber walls in the vertical plane.

The AMX/FMX beamlines for ABBIX project will use an identical configuration to the straight section where the SRX-IVU in the project beamlines is located. The LIX-IVU will be almost identical to a U23L except for the minimum gap (5.7mm) and the vacuum chamber length.

3. Status of the NSLS-II project devices

3.1. Damping Wiggler

The first article damping wiggler was delivered by Danfysik to BNL on November 12, 2012. Various modifications to control system have been implemented to avoid motor stall in certain test modes. Magnetic field measurement has been completed. Figure 1-a and 1-b show a comparison between the vendor's result and BNL's in terms of peak field and period length analysis, respectively. They indicate excellent correspondence between two independent measurements. Other results such as field integrals and multipoles are also satisfactory. The second device has been delivered in June 2013, and the last device is scheduled to arrive at BNL in October 2013.

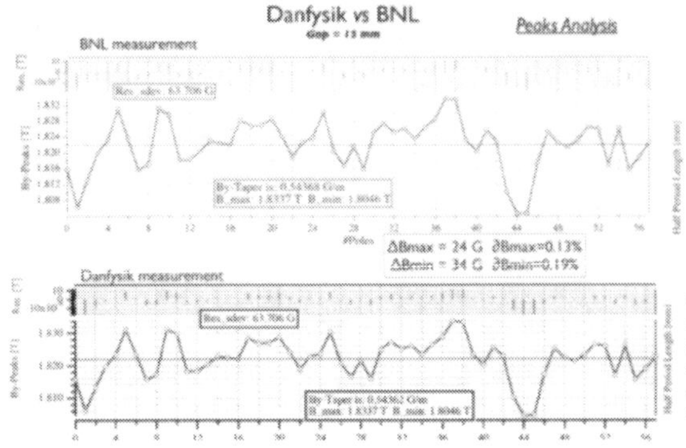

Figure 1-a: Comparison of peak field measurement by the vendor and by BNL.

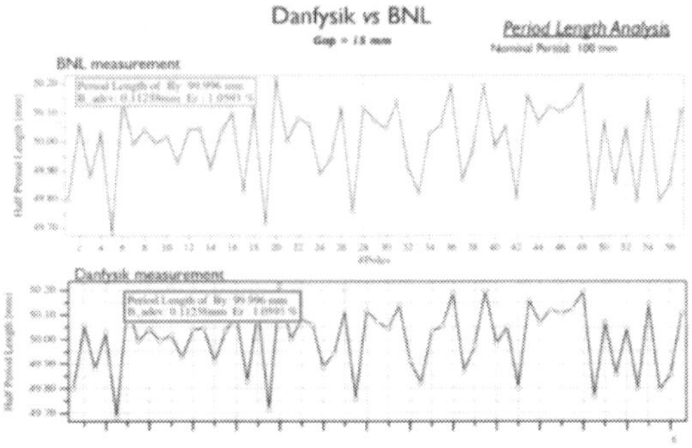

Figure 1-b: Comparison of period length measurement by the vendor and by BNL.

3.2. Elliptically Polarizing Undulator (EPU)

Apple-II type EPUs have been fabricated by Kyma SRL. Vendor field measurement results have been found satisfactory and the devices are almost ready to be shipped. Active field correction scheme by current strips [3] will be implemented on the EPU vacuum chamber. Figure 2 shows a photograph of an EPU at the vendor site and the vacuum chamber including supports is delineated in Fig 3.

Figure 2: One of two EPU49s at the vendor's site.

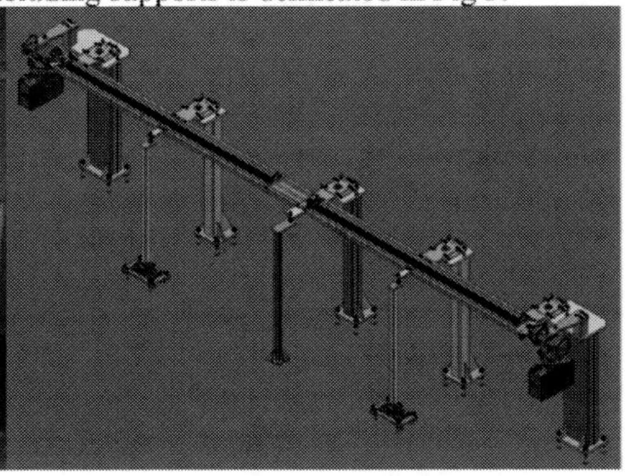

Figure 3: A schematics of CSX-EPU vacuum chamber

3.3. In-Vacuum Undulator (IVU)

Four IVUs are being fabricated by Hitachi Metal America. The first U20 is in its final tuning stage at the SPring-8 Magnetic Measurement Facility (MMF) shown in Fig. 4. The second U20, as well as the first U21 are in the mechanical system construction phase, while a U22 is still in the design stage. Figure 5 is a CAD rendering of a U22 in a long straight section.

Figure 4: The first U20 at the SPring-8 MMF.

Figure 5: CAD rendering of U22 in a long straight section

Journal of Physics: Conference Series **493** (2014) 012031 doi:10.1088/1742-6596/493/1/012031

4. IVMMS

An in-Vacuum Magnetic Measurement System (IVMMS) to be used to measure future 1.5m long CPMUs has been fabricated. Field measurements of PrFeB manget arrays at LN2 temperature have been conducted (Fig.8). A similar result (Fig.7) has been obtained to that taken by our Vertical Test Facility [2].

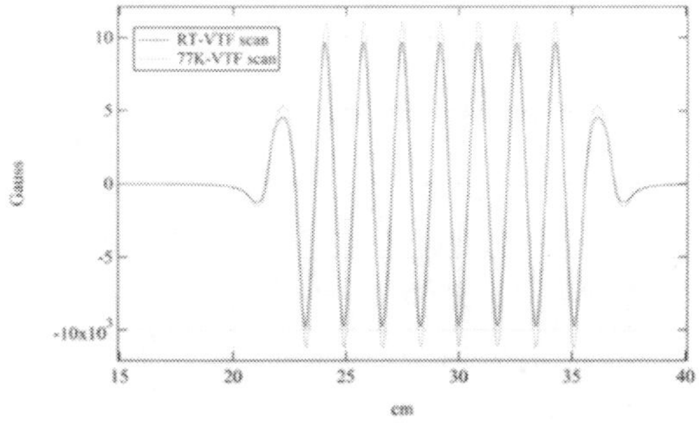

Figure 7: PrFeB Array Measurement by the VTF

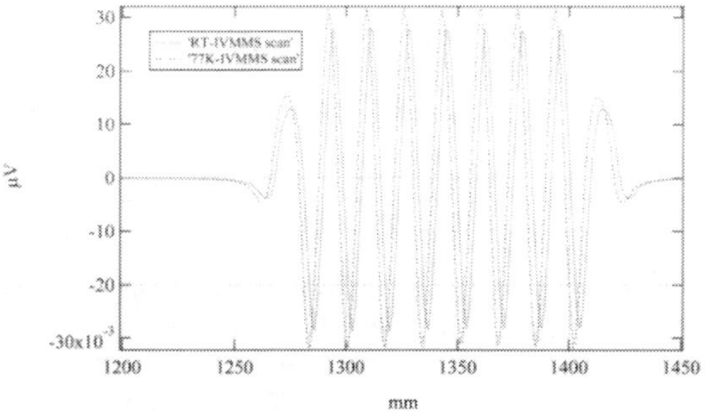

Figure 8: PrFeB Array Measurement by the IVMMS

5. Beam dynamics issues with NEXT-EPUs.

The impact of APPLE-II type EPUs on the storage ring electron beam dynamics has been extensively studied in recent years [3]. Given electron beam parameters in the straight sections where these EPUs are installed, significant impact on the beam dynamics is predicted if no mitigation measures are taken. The main source of this detrimental effect comes from the so-called "dynamic field integral (or multipole) effect". The integrated field seen by sinusoidally wiggling electrons is given by [4]

$$\int B_y ds \approx \frac{-L}{2k^2 B\rho} B_y(x_i)\frac{dB_y(x_i)}{dx} \quad (1)$$

where L is the length of the device, k=2π/period length, ρ is the radius of curvature of the beam trajectory, and x_i is the horizontal displacement of the electron. This effect scales as the ID period length squared and as the derivative of the transverse field roll-off. Field integral measurement capabilities at most insertion device magnetic measurement facilities utilize straight coils and cannot quantify the transverse field roll-off experienced by wiggling e-beam. After vigorous tracking studies, the impact of 7m long SIX-EPUs on the dymamic aperture will not be negligible since they are installed in a long straight section where the vertical beta function is three times bigger than that of a short straight. Searching for a better working point and possible addition of octupole magnets to suppress nonlinearity will be investigated.

[1] http://www.bnl.gov/ps/nsls2/beamlines/

[2] Tanabe, T., et. al.,"Cryo-temperature field measurement of Pr2Fe14B undulator and performance enhancement options at the NSLS-II," The proceedings of the tenth International Conference on Synchrotron Radiation Instrumentation (SRI 2009), September 27 – October 2, 2009, Melbourne, Australia.

[3] J. Bahrdt et al., "Active Shimming of the Dynamic Multipoles of the BESSY UE112 APPLE Undulator" EPAC08, WEPC097 p. 2222.

[4] J. Safranek, et. al., "Nonlinear dynamics in a SPEAR wiggler," Phys. Rev. ST, Acc. and Beam, Vol. 5, p.010701 (2002).

A comprehensive facility for EXAFS measurements at the INDUS-2 synchrotron source at RRCAT, Indore, India

S Basu, C Nayak, A K Yadav, A Agrawal, A K Poswal, D Bhattacharyya[1], S N Jha and N K Sahoo
Applied Spectroscopy Division, Bhabha Atomic Research Centre, Mumbai, Maharashtra, 400085, India

E-mail: dibyendu@barc.gov.in

Abstract. The EXAFS technique, which deals with fine structure oscillations observed in the X-ray absorption spectrum of an element from 50 eV to ~700 eV above its absorption edge, gives precise information regarding the short range order and local structure around the particular atomic species in the material. With the advent of modern bright synchrotron radiation sources, EXAFS has emerged to be the most powerful local structure determination technique, which can be applied to any type of material viz. amorphous, polycrystalline, polymers, surfaces and solutions etc. Over the last few years a comprehensive facility for carrying out EXAFS measurements with synchrotron radiation over variety of samples has been developed at the 2.5 GeV, Synchrotron Radiation Source (INDUS-2) at RRCAT, Indore, India. The facility consists of two operational beamlines viz., the energy dispersive EXAFS beamline (BL-08) and the Energy Scanning EXAFS beamline (BL-09).

1. Introduction

The Extended X-ray Absorption Fine Structure (EXAFS) technique, with synchrotron radiation as the X-ray source, is a powerful tool to study short range order and local structures around any particular element present in a material. X-ray Absorption Fine Structure (XAFS) is the modulation of an atom's X-ray absorption probability due to the chemical and physical state of the atom. XAFS spectra are especially sensitive to the formal oxidation state, coordination chemistry, and the distances, coordination number and species of the neighbouring atoms immediately surrounding the probed element. This technique has the added advantage of being element specific and crystallinity of sample is not a prerequisite. X- rays being fairly penetrating in matter, this technique is not inherently surface-sensitive in the transmission mode and average bulk information from a sample can be obtained. XAFS measurements can even be made on trace amounts of elements giving a unique and direct measurement of chemical and physical state of dilute species in a variety of systems by employing data acquisition in fluorescence mode. Surface sensitivity can be imposed by modifying the measurement technique to total electron yield mode. The experimental X-ray absorption spectrum of a sample consists of two regions: X-ray Absorption Near Edge Structure (XANES) (-50 eV to +50 eV around the absorption edge) and Extended X-ray Absorption Fine Structure (EXAFS) (from 50 eV to 700 – 1000 eV above the absorption edge). The former sheds light on the oxidation state of the central atom in a compound whereas the latter gives relevant as information regarding the bond length

[1]To whom any correspondence should be addressed.

and the coordination number around the probed atom well as disorder factor in the system. The detection of these fine oscillations requires a tunable and bright X-ray source like synchrotron in which radiation is emitted by high energy accelerated electrons. Hence with the advent of modern bright synchrotron radiation sources, this technique has emerged out to be one of the most powerful techniques for local structure determination, which can be applied to any type of material viz. amorphous, polycrystalline, polymers, surfaces and solutions under different ambient and extreme environmental conditions. EXAFS measurement facilities referred to as beamlines, are one of the most significant and extensively used experimental facilities at Synchrotron radiation sources all over the world.

Over the last few years a comprehensive facility for carrying out EXAFS measurements with synchrotron radiation has been developed at the INDUS-2 Synchrotron Radiation Source at RRCAT, Indore, India. INDUS-2 is a 2.5 GeV 3rd generation machine, presently working at 120 mA beam current (designed for 300 mA) with a horizontal beam emittance of 3.9×10^{-8} mrad and a vertical beam emittance of 3.9×10^{-9} mrad. The beam stability is good with a noise level of ~ 2%. The EXAFS facility consists of two operational beamlines built at bending magnet ports of INDUS-2 viz., the Energy Dispersive EXAFS beamline (BL-08) and the Energy Scanning EXAFS beamline (BL-09) [1].

2. Description of EXAFS facilities

2.1 Energy Dispersive EXAFS beamline (BL-08)

EXAFS measurements with synchrotron radiation are generally carried out in two different modes, viz. energy dispersive mode and energy scanning mode. In the dispersive EXAFS mode of measurement, a bent crystal polychromator is used to select a band of energy from the white synchrotron beam horizontally dispersed and focused on the sample. The transmitted beam intensity from the sample is recorded on a position-sensitive CCD detector, recording the whole EXAFS spectrum around an absorption edge in a single shot. This is essentially the working principle of the energy dispersive EXAFS beamline (BL-08) at the INDUS-2 which uses a 460 mm long Si (111) single crystal mounted on a mechanical crystal bender which can bend the crystal to the shape of an elliptical cylinder in such a way that the source and the sample position are at the two foci of an ellipse as evident in figure 1. [2]. The crystal selects a particular band of energy from white synchrotron radiation depending on the grazing angle of incidence of the synchrotron beam (Bragg angle) and disperses as well as focuses the band on the sample.

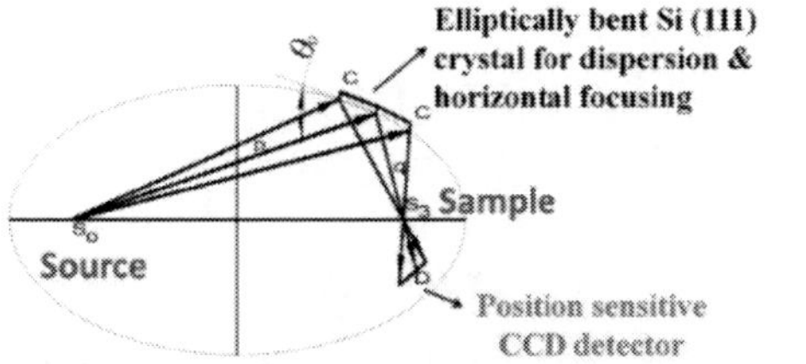

Figure 1. Optical geometry of BL-08

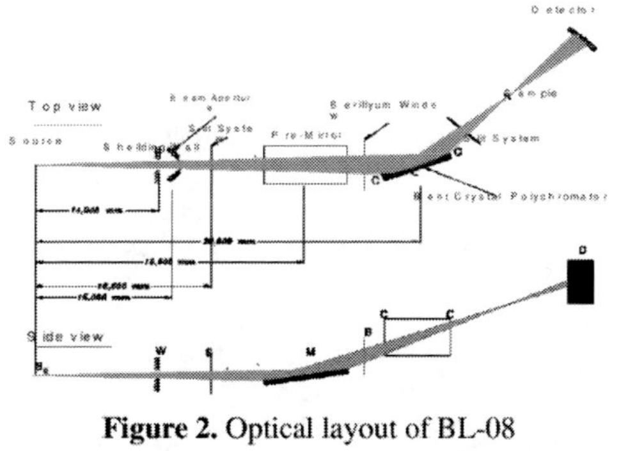

Figure 2. Optical layout of BL-08

A Rh coated cylindrical pre-mirror with meridional curvature is used for rejection of higher harmonics and vertical focusing of the beam. The radiation transmitted through the sample is detected by a position sensitive CCD detector having 2048 x 2048 pixels. The whole absorption spectrum of the sample can be recorded simultaneously on the detector within fraction of a second. The optical layout of the beamline is depicted in figure 2. The beamline is designed to cover the photon energy range of 5-20 keV providing energy bandwidths of 0.3, 1.0 and 2.0 keV and with resolution of 0.5, 1 and 2 eV per pixel at photon energies of 5, 10 and 20 keV, respectively. The beamline is particularly useful for in-situ and time-resolved studies on samples in transmission geometry.

2.2 Energy Scanning EXAFS beamline (BL-09)

This is the more conventional mode of EXAFS measurement in which the beamline uses a double-crystal monochromator (DCM) to select a particular energy from the incoming synchrotron beam that is incident on the sample. The intensity of the transmitted beam passing through the sample or that of the fluorescence beam emerging out of the sample is recorded along with the incident intensity at each energy. The absorption data of the sample are then recorded by scanning the monochromator over the whole energy range of interest. The Energy Scanning EXAFS beamline (BL-09), which has been commissioned recently, covers a photon energy range of 4-25 KeV and has a resolution ($\Delta E/E$) of 10^{-4} at 10 KeV. The optical layout of the beamline is shown in figure 3. The beamline uses a 1.3 m long

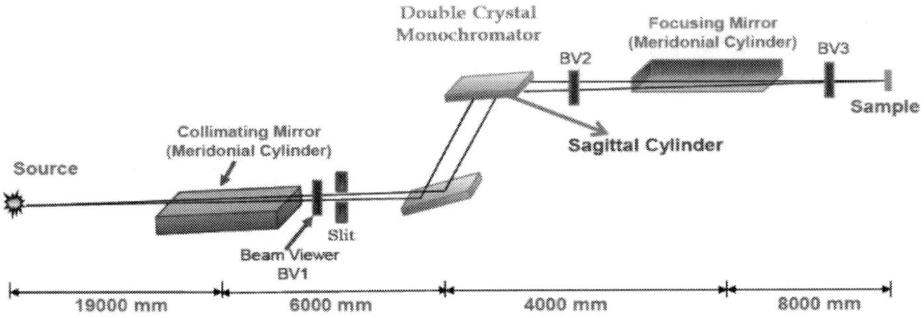

Figure 3. Optical Layout of BL-09

meridional cylindrical mirror with **a** radius of curvature **of** 11.2 Kms coated with Rh/Pt as a collimator. This mirror is followed by a double crystal monochromator (DCM). The DCM consists of two Si (111) crystals with 2d = 6.2709Å and narrow Darwin width (~ 6 arc sec). 1st crystal is cooled by indirect water-cooled arrangement and temperature is stable within ± 0.1°C. The second crystal of the DCM is a sagittal cylinder with radius of curvature in the range 1.28-12.91 m which provides horizontal focusing to the beam. The impact of the bending of the 2nd crystal on the over-all resolution of the monochromator was investigated by carrying out high resolution measurements of pre-edge peaks of transition metal elements viz., V, Ti, Mn etc. where no significant change in resolution was observed. A post mirror with meridional curvature is used upside-down for bending the beam path to horizontal

Table 1. Beamline Specifications

	BL-08	**BL-09**
Energy range	5-20 keV	4-25 keV
Energy Resolution (E/ΔE)	10^4 @ 10keV	10^4 @ 10keV
Flux@ Focus	10^{12} photons/sec/1000 eV bandwidth @2.5 GeV, 300mA	10^{11} ph/sec/0.1% band width @2.5 GeV, 300mA
Focal spot size	0.20 mm x 0.40 mm	0.50 mm x 0.50 mm

direction again as well as for vertical focusing of the beam at the sample position. For measurements in the transmission mode, the sample is placed between two ionization chamber detectors. The first ionization chamber measures the incident flux (I_0) and the second ionization chamber measures the transmitted intensity (I_t). A third ionization chamber is also used at the end for measuring spectra of reference foils to carry out corrections in energy calibration of the monochromator. It is also possible to carry out measurement in fluorescence mode and a Lytle-type ionization chamber based detector or a Si drift solid state detector is used for this purpose which is placed in front of the sample in a 45° geometry. This beamline is useful for measurement on trace elements, surface studies and thin film materials on a thick substrate.

2.3 Sample environments

In the above two beamlines experiments can be performed in different sample environments. These include a low temperature He-gas cooled closed cycle cryostat which can go down to 4.2 K and a high temperature cell that can go up to 1000K. In the high temperature cell it is possible to carry out measurements under various gaseous environments viz. Cl, H_2S, inert gases etc. The sample can also be handled in different forms such as powdered solid compressed to pellet form and sample on tape. A special quartz cell fitted with Be window is also available for measurement of liquid samples. An ellipsoidal mirror for focussing of the beam to <100μm at the sample position is presently being installed at BL-08 to enable measurements under high pressure employing diamond anvil cells. X-ray Magnetic Circular Dichroism (XMCD), for high Z elements utilizing the polarizing property of the synchrotron radiation, and measurements on radioactive samples inside a specially fabricated glove box are planned for beamline upgrades that will augment future programs.

3. Conclusion

An EXAFS measurement facility consisting of two beamlines working in energy dispersive mode and energy scanning mode has been developed and commissioned respectively at BL-08 and BL-09 bending magnet ports of INDUS-2 Synchrotron source at the Raja Ramanna Centre for Advanced Technology (RRCAT), Indore, India. The energy dispersive EXAFS beamline is suitable for time resolved studies in transmission mode while in the energy scanning beamline measurement is possible both in transmission and fluorescence mode. Thus these two beamlines are complementary to each other and together they offer a comprehensive EXAFS measurement facility that cover a wide range of samples including dilute materials, thin films, chemical catalysts, liquid etc. This indigenously developed facility is regularly used by researchers from different national laboratories and universities of India and has resulted in a number of research papers in peer reviewed journals. Some of the measurements carried out include studies of Gd, Nd and La doped ZrO_2 systems, which have potential uses as electrolyte materials for solid oxide fuel cells [3,4]; studies of Eu doped $SrSnO_3$ nanoparticles [5] and Eu^{3+} doped GaOOH nanorods for their luminescence properties [6]; studies of Sb doped Bi_2UO_6, which is a potential by-product in high temperature nuclear reactors [7]; and studies of Fe doped $CaMnO_3$ used as a magnetic material [8].

References

[1] Website: http://www.cat.gov.in/technology/accel/srul/beamlines/index.html
[2] Bhattacharyya D, Poswal A K, Jha S N, Sangeeta and Sabharwal S C 2009 *Nuclear Instruments andMethods in Physics Research A* **609** 286.
[3] Basu S, Varma Salil, Shirsat A N, Wani B N, Bharadwaj S R, Chakrabarti Aparna, Jha S N and Bhattacharrya D 2012 *J. Appl. Phys* **111** 053532.
[4] Basu S, Varma Salil, Shirsat A N, Wani B N, Bharadwaj S R, Chakrabarti Aparna, Jha S N and Bhattacharyya D 2013 *J. Appl. Phys.* **113** 043508.
[5] Basu S, Patel D K, Nuwad J, Sudarsan V, Jha S N, Bhattacharyya D, Vatsa R K and Kulsreshtha S K 2013 *Chem. Phys. Lett.* **561–562** 82.
[6] Basu S, Naidu B S, Pandey M, Sudarsan V, Jha S N, Bhattacharyya D, Vatsa R K and Kshirsagar R J 2012 *Chem. Phys. Lett.* **528** 21.
[7] Misra N L, Yadav A K, Dhara Sangita, Phatak Rohan, Mishra S K, Poswal A K, Jha S N, Sinha A K and Bhattacharyya D 2013 *Analytical Sciences* **29** 579.
[8] Neetika, Das A, Dhiman I, Nigam A K, Yadav A K, Bhattacharyya D and Meena S S 2012 *J. Appl. Phys.* **112** 123913.

Journal of Physics: Conference Series **493** (2014) 012033 doi:10.1088/1742-6596/493/1/012033

X-ray Emission Collected in a Novel Energy Dispersive Approach

KD Finkelstein[1], D Agyeman-Budu[1], A Lyndaker[1], C Pollock[2], T Krawczyk[1]
[1]CHESS Cornell University, [2]Max-Plank Institute fur Chemische Energiekonversion

Email: kdf1@cornell.edu

Abstract. Novel methods & materials have been used to produce dynamically bent Bragg diffraction analyzer crystals in a modified von Hamos geometry.They are used to collect a wide energy range X-ray emission spectrum over a large solid angle with electron volt resolution. Crystals fabricated from silicon-on-insulator wafers by photolithography and deep reactive ion etching can bend to 10 cm radius without increased lattice strain. The design permits adjustment of energy dispersion for individual analyzers in an array. A multilayer mono, mono-capillary focusing, and multi-crystal spectrometer together collect signals at a bend magnet beamline comparable to those from an undulator. Preliminary measurements validate this new energy dispersive spectrometer.

1. Introduction

X-ray emission spectroscopy (XES) is one of several photon-in/photon-out methods to probe electronic structure with element selectivity [1]. Here we consider non-resonant emission where the incident energy is set well above the absorption edge and the emitted x-ray fluorescence is collected at eV resolution. The energy dispersive spectrometer we describe is appropriate for current XES studies and will help to enable time-dependent studies that require 4th generation, ultra-bright sources. Bragg reflection spectrometers, based on flat crystals, von Hamos analyzers [2], can simultaneously collect emission spectra over 100eV at 10KeV. A significant challenge with such analyzers is low signal rate since flat crystals intercept a small solid angle per unit energy interval and reflect a bandwidth typically a small fraction of this interval.

Under conditions in Figure 1, the spectrometer resolution is $dE/E|_{spect} = \cot\Theta_B\,(\delta/2R)\,\sin^2\Theta_B$, with detector pixel size δ, Bragg angle Θ, and bend radius R. Reflection bandwidth is $dE/E|_{ana} = \cot\Theta_B\,\Delta\Theta_{Darwin}$, and solid angle intercepted by a pixel is $d\Omega = \varphi\delta\sin^3\Theta\,/2R$ where φ=RS is the angle subtended bending analyzer of length S about the joining line. Signal increases with decreasing R and as $\varphi\rightarrow 2\pi$. The spectrometer reported here significantly reduces R without increasing crystal strain, increases signal using multiple crystals adjusted to a common dispersion (energy vs. position) at the detector, and uses novel upstream optics to increase flux on the specimen.

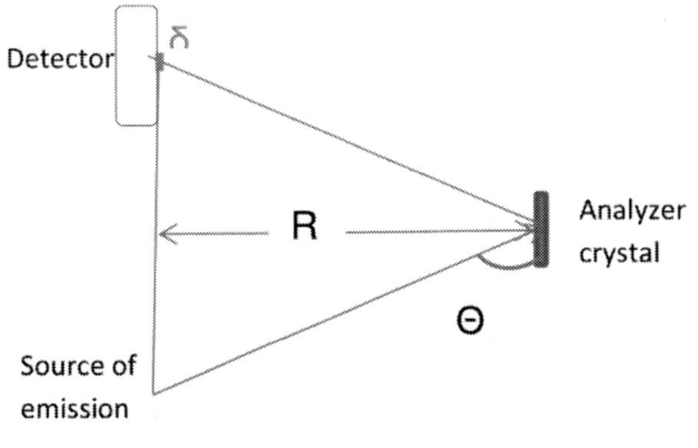

Figure. 1: The plane of diffraction of a von Hamos spectrometer with a point source of emission. The signal is increased by bending the analyzer around the "joining line" between source & detector.

2. Design & fabrication of bendable analyzer crystals

Deep reactive ion etching (dRIE) for fabricating bent silicon x-ray optics was first reported with a sagittal focusing monochromator [3]. Etch depth varies 5% over the 100 mm wafer area and will distort curved optics, so we suggested using bonded (bilayer) wafers [4]. This was realized using silicon-on-insulator (SOI) [5] with a thick FZ silicon diffraction layer bonded to a thin CZ silicon bendable layer through a micron thick SiO_2 layer. The FZ surface is lithographically patterned to form the dRIE etch mask. The "bar analyzer" shown mounted in Figure 2, consists of narrow strips defined by trenches in the FZ layer; only the CZ layer bends. Strips have length (in diffraction plane) H and width w, trench width is g, and analyzer active width is S.

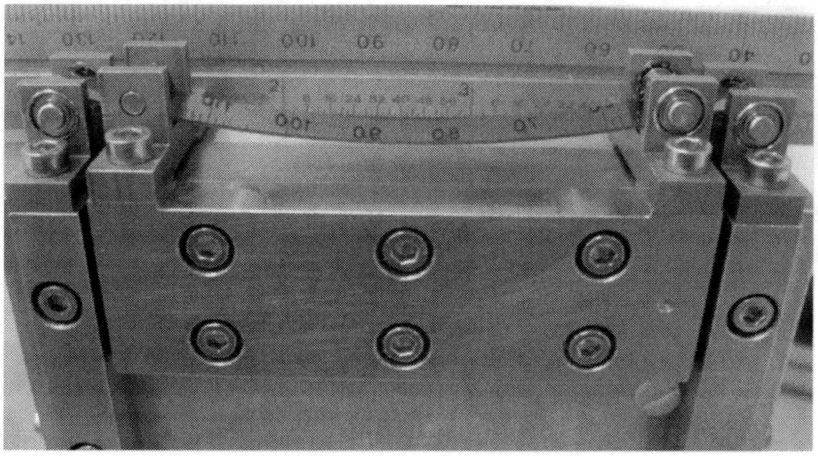

Figure 2: "Bar analyzer" bent in a simple bending apparatus. The diffraction plane contains the bend axis and is normal to crystal surface. Bending occurs when the pair of inner rods move down relative to outer rods. All 4 rods are free to rotate.

The energy band collected depends on Θ_B, R, H as $dE/E|_{collected} = H\sin(2\Theta_B)/2R$. Focus size (transverse to dispersion direction) at detector is 2w. $S \approx Nw$, with N the number of strips across the bar. A simple beam bending model for trench width g, illustrated in Figure 3, assumes uniform bending confined to layer thickness t where local bend radius $r = Rg/(w+g) \geq 0.5Et/\sigma$. If silicon has elastic modulus E= 190GPa, and maximum stress σ= 2GPa, then r= 6mm with safety factor 3 for t = 40μm. So analyzers with w=500μm, g = 32μm will bend to R < 100mm.

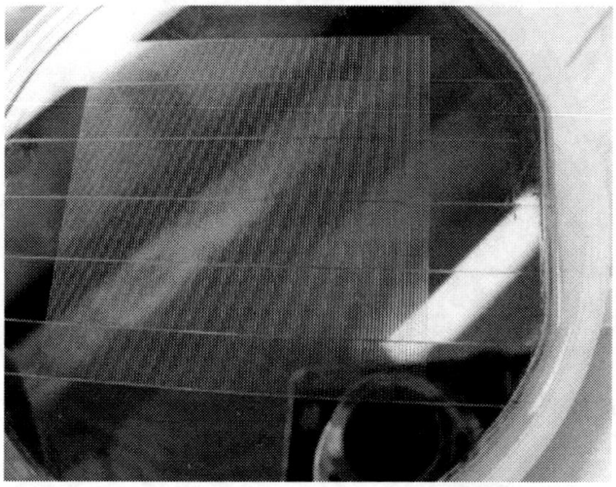

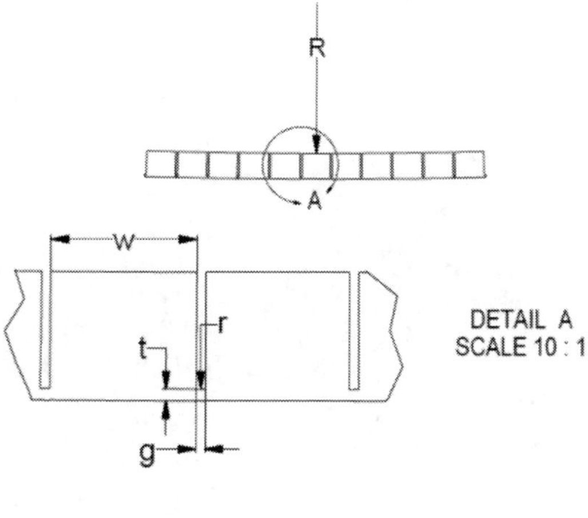

Figure 3: left) 100mm diameter silicon wafer, with bar analyzer pattern produced by deep-reactive ion etching. right) Model used to estimate "bar analyzer" design parameters: w, t, g.

Journal of Physics: Conference Series **493** (2014) 012033 doi:10.1088/1742-6596/493/1/012033

3. Superimposing signals

The first measurements collected were copper Kβ emission using silicon (553) von Hamos analyzers. Collecting weak Kβ emission requires intercepting a large solid angle. This is done by increasing the number of analyzers and decreasing R. A multiplexing scheme to match dispersion of energy with position on the detector, for 3 analyzers, is illustrated in Figure 4. Analyzers are oriented so beam incidence angles match at corresponding points on each surface, and diffracted rays of common energy intersect on the detector. This occurs when distance from source to center of analyzer n is $L(\varphi_n) = R\sin(\Theta_0 + \varphi_n)/\sin^2\Theta_0$, and if the detector face is tilted away from joining line by $90 - \Theta_0$. Θ_0 is Bragg angle center and $\varphi_n > 0$ for crystals positioned above center crystal relative to source point. Figure is drawn for center plane where crystals are furthest from the joining line. To achieve best resolution analyzers should bend so $R_n = L_n \sin\Theta_0$ where each energy diffracts along circular arcs of length S. In this case analyzers with $\varphi \neq 0$ focus slightly off the detector plane, therefore focusing on this plane produces very slight dispersion errors.

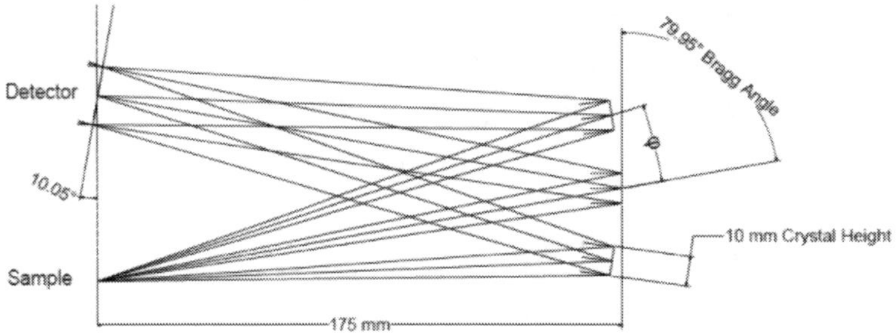

Figure 4: Configuration for collecting copper Kβ emission using silicon (553) reflection. Corresponding incident and diffracted rays shown for each crystal; center crystal is vertical.

4. Realized spectrometer

The spectrometer in Figure 5 was tested collecting copper Kβ emission with 3 Si(553) analyzers ($\Theta_B = 79.95°$) with R_{center} =175mm, S ≈ 55mm, using Dectris100K detector with 172μm square pixels. For H=1cm, $dE|_{collected}$ = 87 eV with dΩ =147 μSrad/analyzer/1 eV. Each analyzer is: bent to adjust R, rotated to adjust energy, & translated to adjust dispersion. Translating the spectrometer parallel to incident beam changes Bragg angle Θ and 2Θ changes as the detector is also translated. The 10 KeV incident beam, with bandwidth 1.2%, was focused to 140 μm on a pressed powder sample using a 125 mm working distance monocapillary built at CHESS.

Figure 5: CHESS multi-crystal spectrometer with detector (on black rotation stage) and sample (on copper mount) removed for clarity. Downstream is to left. Center crystal (#3) is oriented parallel to line joining detector and sample centers. Focusing capillary is on alignment stage upstream of sample.

5. Results

This work demonstrates how novel, highly flexible analyzers can be combined to multiply signal collection rate and match energy dispersion to overlap spectra from multiple crystals without reducing resolution. The dispersion matching works, but our benders introduce displacement of spectra transverse to the dispersion direction. In practice this problem actually simplifies dispersion adjustment. Figure 6 shows the detector image and resultant spectrum from a single analyzer. Figure 7 (left) shows $K\beta_{1,3}$ peaks from three analyzers and (right) scaled to common intensity, one sees agreement in line shape and centroid resulting from dispersion match. These spectra would be summed to increase statistical precision before further analysis.

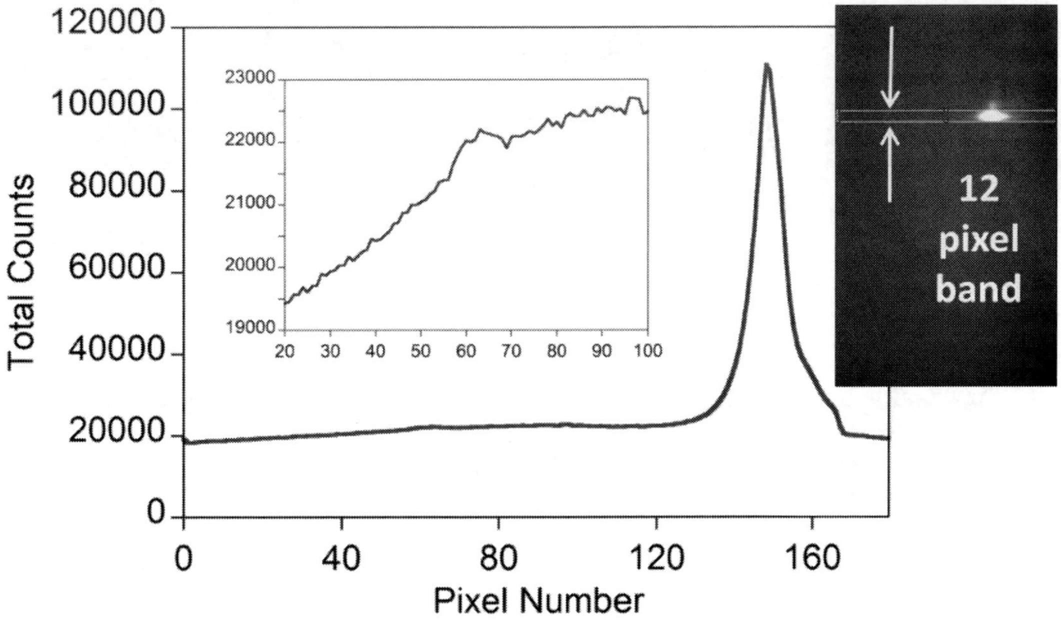

Figure 6: Copper $K\beta$ spectrum from CuO2 collected by single analyzer; energy decreases with pixel number. Analyzer R=175mm, collection time 3000 sec. Inserts: (upper left) valence-to-core (V2C) region, (right) detector image. Spectrum is vertical sum over 12 pixels. Background subtracted peak count rate is 30Hz and $K\beta_{1,3}$ is 8.65 pixels wide, so line is collected at 277Hz. Known separation between V2C and $K\beta_{1,3}$ calibrates the horizontal axis at 0.82eV/pixel and yields $K\beta_{1,3}$ peak width 7.06eV, in agreement with Reference 6. The large background, from close proximity of sample, analyzer and detector, precludes quantitative analysis of V2C line shape.

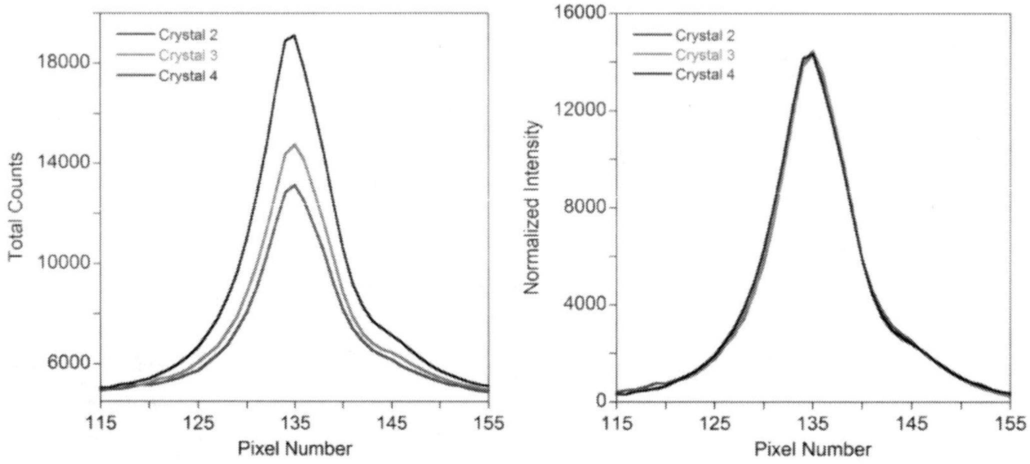

Figure 7: To adjust energy dispersion and peak position, analyzer crystals are translated in spectrometer center plane. Plot at left shows spectral peaks from 3 crystals. In the plot at right individual spectra have been background subtracted and peak intensities have been scaled to show agreement in line shape and position.

6. Conclusion and directions of continued development

Results are reported on the development of a multi-crystal von Hamos analyzer that multiplies rate of signal collection for x-ray emission with eV resolution. The novel method of analyzer fabrication produces highly flexible analyzers and the new arrangement of crystals demonstrates matching of energy dispersion at the detector. The prototype spectrometer produces dynamical bending using 4-pin benders with limited mechanical adjustment that produces image aberrations. We conclude that dynamic bending is very hard to realize in the compact size required to maximize collection solid angle/analyzer, so a 2nd generation spectrometer with fixed bend radius and optimum collection capability is now being developed.

References

[1] Overview and references may be found in Synchrotron Radiation News Vol.22, 2, 12-16 (2009).

[2] J. Hoszowska, J.-Cl. Dousse, J. Kern and Ch. Rhême, *High resolution Von Hamos Crystal x-ray Spectrometer,* Nucl. Instrum. Meth. A **376**, 129 (1996).

[3] K.D.Finkelstein, S.Rosenblatt, P.Cottle, *Rev.of Sci.Instr.*,73,1464 (2002).

[4] K.D. Finkelstein, P. Abbamonte, V. Kostroun, Proceedings of SPIE 4783 (2002).

[5] Celler, G. K.; Cristoloveanu, S. (2003). "Frontiers of silicon-on-insulator". *J Appl Phys* **93** (9): 4955.
We use SOI made by Ultrasil Corporation in Hayward, Calif. USA http://www.ultrasil.com/

[6] J.Szlachetko, M.Nachtegaal, E.de Boni, M.Willimann, O.Safonova, J.Sa, G.Smolentsev, M.Szlachetko, J.A.van Bokhoven,J-Cl.Dousse,J.Hoszowska,Y.Kayser,P.Jagodzinski,A.Bergamaschi,B.Schmitt,C.David, and A.Lücke; REVIEW OF SCIENTIFIC INSTRUMENTS **83**, 103105 (2012).

Institute of Physics
Dirac House, Temple Back
Bristol BS1 6BE UK

CURRAN ASSOCIATES INC
proceedings.com